Learn R

Chapman & Hall/CRC
The R Series

Series Editors
John M. Chambers, Department of Statistics, Stanford University, California, USA
Torsten Hothorn, Division of Biostatistics, University of Zurich, Switzerland
Duncan Temple Lang, Department of Statistics, University of California, Davis, USA
Hadley Wickham, RStudio, Boston, Massachusetts, USA

*** Recently Published Titles***

Geocomputation with R
Robin Lovelace, Jakub Nowosad, Jannes Muenchow

Advanced R, Second Edition
Hadley Wickham

Dose Response Analysis Using R
Christian Ritz, Signe Marie Jensen, Daniel Gerhard, Jens Carl Streibig

Distributions for Modelling Location, Scale, and Shape
Using GAMLSS in R
Robert A. Rigby , Mikis D. Stasinopoulos, Gillian Z. Heller and Fernanda De Bastiani

Hands-On Machine Learning with R
Bradley Boehmke and Brandon Greenwell

Statistical Inference via Data Science
A ModernDive into R and the Tidyverse
Chester Ismay and Albert Y. Kim

Reproducible Research with R and RStudio, Third Edition
Christopher Gandrud

Interactive Web-Based Data Visualization with R, plotly, and shiny
Carson Sievert

Learn R
Pedro J. Aphalo

For more information about this series, please visit: https://www.crcpress.com/Chapman--HallCRC-The-R-Series/book-series/CRCTHERSER

Learn R
As a Language

Pedro J. Aphalo

CRC Press
Taylor & Francis Group
Boca Raton London New York

CRC Press is an imprint of the
Taylor & Francis Group, an **informa** business

A CHAPMAN & HALL BOOK

First edition published 2020
by CRC Press
6000 Broken Sound Parkway NW, Suite 300, Boca Raton, FL 33487-2742

and by CRC Press
2 Park Square, Milton Park, Abingdon, Oxon, OX14 4RN

© 2020 Taylor & Francis Group, LLC

CRC Press is an imprint of Taylor & Francis Group, LLC

ISBN: 978-0-367-18253-3 (pbk)
ISBN: 978-0-367-18255-7 (hbk)
ISBN: 978-0-429-06034-2 (ebk)

Contents

Preface

> "Suppose that you want to teach the 'cat' concept to a very young child. Do you explain that a cat is a relatively small, primarily carnivorous mammal with retractible claws, a distinctive sonic output, etc.? I'll bet not. You probably show the kid a lot of different cats, saying 'kitty' each time, until it gets the idea. To put it more generally, generalizations are best made by abstraction from experience."
>
> R. P. Boas
> *Can we make mathematics intelligible?*, 1981

This book covers different aspects of the use of the R language. Chapters 1 to 5 describe the R language itself. Later chapters describe extensions to the R language available through contributed *packages*, the *grammar of data* and the *grammar of graphics*. In this book, explanations are concise but contain pointers to additional sources of information, so as to encourage the development of a routine of independent exploration. This is not an arbitrary decision, this is the normal *modus operandi* of most of us who use R regularly for a variety of different problems. Some have called approaches like the one used here "learning the hard way," but I would call it "learning to be independent."

I do not discuss statistics or data analysis methods in this book; I describe R as a language for data manipulation and display. The idea is for you to learn the R language in a way comparable to how children learn a language: they work out what the rules are, simply by listening to people speak and trying to utter what they want to tell their parents. Of course, small children receive some guidance, but they are not taught a prescriptive set of rules like when learning a second language at school. Instead of listening, you will read code, and instead of speaking, you will try to execute R code statements on a computer—i.e., you will try your hand at using R to tell a computer what you want it to compute. I do provide explanations and guidance, but the idea of this book is for you to use the numerous examples to find out by yourself the overall patterns and coding philosophy behind the R language. Instead of parents being the sound board for your first utterances in R, the computer will play this role. You will *play* by modifying the examples, see how the computer responds: does R understand you or not? Using a language actively is the most efficient way of learning it. By using it, I mean actually reading, writing, and running scripts or programs (copying and pasting, or typing ready-made examples from books or the internet, does not qualify as using a language).

I have been using R since around 1998 or 1999, but I am still constantly learning new things about R itself and R packages. With time, it has replaced in my work as a researcher and teacher several other pieces of software: SPSS, Systat, Origin, MS-Excel, and it has become a central piece of the tool set I use for producing lecture slides, notes, books, and even web pages. This is to say that it is the most useful piece of software and programming language I have ever learned to use. Of course, in time it will be replaced by something better, but at the moment it is a key language to learn for anybody with a need to analyze and display data.

What is a language? A language is a system of communication. R as a language allows us to communicate with other members of the R community, and with computers. As with all languages in active use, R evolves. New "words" and new "constructs" are incorporated into the language, and some earlier frequently used ones are relegated to the fringes of the corpus. I describe current usage and "modisms" of the R language in a way accessible to a readership unfamiliar with computer science but with some background in data analysis as used in biology, engineering, or the humanities.

When teaching, I tend to lean toward challenging students, rather than telling an over-simplified story. There are two reasons for this. First, I prefer as a student, and I learn best myself, if the going is not too easy. Second, if I would hide the tricky bits of the R language, it would make the reader's life much more difficult later on. You will not remember all the details; nobody could. However, you most likely will remember or develop a sense of when you need to be careful or should check the details. So, I will expose you not only to the usual cases, but also to several exceptions and counterintuitive features of the language, which I have highlighted with icons. Reading this book will be about exploring a new world; this book aims to be a travel guide, but neither a traveler's account, nor a cookbook of R recipes.

Keep in mind that it is impossible to remember everything about R! The R language, in a broad sense, is vast because its capabilities can be expanded with independently developed packages. Learning to use R consists of learning the basics plus developing the skill of finding your way in R and its documentation. In early 2020, the number of packages available in the Comprehensive R Archive Network (CRAN) broke the 15,000 barrier. CRAN is the most important, but not only, public repository for R packages. How good a command of the R language and packages a user needs depends on the type of activities to be carried out. This book attempts to train you in the use of the R language itself, and of popular R language extensions for data manipulation and graphical display. Given the availability of numerous books on statistical analysis with R, in the present book I will cover only the bare minimum of this subject. The same is true for package development in R. This book is somewhere in-between, aiming at teaching programming in the small: the use of R to automate the drudgery of data manipulation, including the different steps spanning from data input and exploration to the production of publication-quality illustrations.

As with all "rich" languages, there are many different ways of doing things in R. In almost all cases there is no one-size-fits-all solution to a problem. There is always a compromise involved, usually between time spent by the user and processing time required in the computer. Many of the packages that are most popular nowadays did not exist when I started using R, and many of these packages make new approaches available. One could write many different R books with a given aim using substantially different ways of achieving the same results. In this book, I limit myself to packages that are currently popular and/or that I consider elegantly designed. I have in particular tried to limit myself to packages with similar design philosophies, especially in relation to their interfaces. What is elegant design, and in particular what is a friendly user interface, depends strongly on each user's preferences and previous experience. Consequently, the contents of the book are strongly biased by my own preferences. I have tried to write examples in ways that execute fast without compromising readability. I encourage readers to take this book as a starting point for exploring the very many packages, styles, and approaches which I have not described.

I will appreciate suggestions for further examples, and notification of errors and unclear sections. Because the examples here have been collected from diverse sources over many years, not all sources are acknowledged. If you recognize any example as yours or someone else's, please let me know so that I can add a proper acknowledgement. I warmly thank the students who have asked the questions and posed the problems that have helped me write this text and correct the mistakes and voids of previous versions. I have also received help on online forums and in person from numerous people, learned from archived e-mail list messages, blog posts, books, articles, tutorials, webinars, and by struggling to solve some new problems on my own. In many ways this text owes much more to people who are not authors than to myself. However, as I am the one who has written this version and decided what to include and exclude, as author, I take full responsibility for any errors and inaccuracies.

Why have I chosen the title "*Learn R: As a Language*"? This book is based on exploration and practice that aims at teaching to express various generic operations on data using the R language. It focuses on the language, rather than on specific types of data analysis, and exposes the reader to current usage and does not spare the quirks of the language. When we use our native language in everyday life, we do not think about grammar rules or sentence structure, except for the trickier or unfamiliar situations. My aim is for this book to help you grow to use R in this same way, to become fluent in R. The book is structured around the elements of languages with chapter titles that highlight the parallels between natural languages like English and the R language.

I encourage you to approach R like a child approaches his or her mother tongue when first learning to speak: do not struggle, just play, and fool around with R! If the going gets difficult and frustrating, take a break! If you get a new insight, take a break to enjoy the victory!

Acknowledgements

First I thank Jaakko Heinonen for introducing me to the then new R. Along the way many well known and not so famous experts have answered my questions in usenet and more recently in Stackoverflow. As time went by, answering other people's questions, both in the internet and in person, became the driving force for me to delve into the depths of the R language. Of course, I still get stuck from time to time and ask for help. I wish to warmly thank all the people I have interacted with in relation to R, including members of my own research group, students participating in the courses I have taught, colleagues I have collaborated with, authors of the books I have read and people I have only met online or at conferences. All of them have made it possible for me to write this book. This has been a time consuming endeavour which has kept me too many hours away from my family, so I specially thank Tarja, Rosa and Tomás for their understanding. I am indebted to Tarja Lehto, Titta Kotilainen, Tautvydas Zalnierius, Fang Wang, Yan Yan, Neha Rai, Markus Laurel, other colleagues, students and anonymous reviewers for many very helpful comments on different versions of the book manuscript, Rob Calver, as editor, for his encouragement and patience during the whole duration of this book writing project, Lara Spieker, Vaishali Singh, and Paul Boyd for their help with different aspects of this project.

Icons used to mark different content

Text boxes are used throughout the book to highlight content that plays specific roles in the learning process or that require special attention from the reader. Each box contains one of five different icons that indicate the type of its contents as described below.

Signals *playground* boxes which contain open-ended exercises—ideas and pieces of R code to play with at the R console.

Signals *advanced playground* boxes which will require more time to play with before grasping concepts than regular *playground* boxes.

Signals important bits of information that must be remembered when using R—i.e., explain some unusual feature of the language.

Signals in-depth explanations of specific points that may require you to spend time thinking, which in general can be skipped on first reading, but to which you should return at a later peaceful time, preferably with a cup of coffee or tea.

Signals text boxes providing general information not directly related to the R language.

1

R: The language and the program

> In a world of ... relentless pressure for more of everything, one can lose sight of the basic principles—simplicity, clarity, generality—that form the bedrock of good software.
>
> Brian W. Kernighan and Rob Pike
> *The Practice of Programming*, 1999

1.1 Aims of this chapter

In this chapter you will learn some facts about the history and design aims behind the R language, its implementation in the R program, and how it is used in actual practice when sitting at a computer. You will learn the difference between typing commands interactively, reading each partial response from R on the screen as you type versus using R scripts to execute a "job" which saves results for later inspection by the user.

I will describe the advantages and disadvantages of textual command languages such as R compared to menu-driven user interfaces as frequently used in other statistics software and occasionally also with R. I will discuss the role of textual languages in the very important question of reproducibility of data analyses.

Finally you will learn about the different types and sources of help available to R users, and how to best make use of them.

1.2 R

1.2.1 What is R?

Most people think of R as a computer program. R is indeed a computer program— a piece of software— but it is also a computer language, implemented in the R program. Does this make a difference? Yes. Until recently we had only one mainstream implementation of R, the program R. Recently another implementation has gained some popularity, Microsoft R Open (MRO), which is directly based on the R program from *The R Project for Statistical Computing*. MRO is described as an enhanced distribution of R. These two very similar implementations are not the only ones available, but others are not in widespread use. In other words, the R language can be used not only in the R program, and it is feasible that other implementations will be developed in the future.

The name "base R " is used to distinguish R itself, as in the R distribution, from R in a broader sense, which includes independently developed extensions that can be loaded from separately distributed extension packages.

Being that R is essentially a command-line application, it can be used on what nowadays are frugal computing resources, equivalent to a personal computer of three decades ago. R can run even on the Raspberry Pi, a micro-controller board with the processing power of a modest smart phone. At the other end of the spectrum, on really powerful servers, R can be used for the analysis of big data sets with millions of observations. How powerful a computer you will need will depend on the size of the data sets you want to analyze, on how patient you are, and on your ability to write "good" code.

One could think of R as a dialect of an earlier language, called S. S evolved into S-Plus (Becker et al. 1988). S and S-Plus are commercial programs, and variations in the language appeared only between versions. R started as a poor man's home-brewed implementation of S, for use in teaching. Initially R, the program, implemented a subset of the S language. The R program evolved until only relatively few differences between S and R remained, and these differences are intentional— thought of as significant improvements. As R overtook S-Plus in popularity, some of the new features in R made their way back into S-Plus. R is free and open-source and the name Gnu S is sometimes used to refer to R.

What makes R different from SPSS, SAS, etc., is that S was designed from the start as a computer programming language. This may look unimportant for someone not actually needing or willing to write software for data analysis. However, in reality it makes a huge difference because R is easily extensible. By this we mean that new functionality can be easily added, and shared, and this new functionality is to the user indistinguishable from that built into R. In other words, instead of having to switch between different pieces of software to do different types of analyses or plots, one can usually find an R package that will provide the tools to do the job within R. For those routinely doing similar analyses the ability to write a short program, sometimes just a handful of lines of code, allows automation of routine analyses. For those willing to spend time programming, they have the door open to building the tools they need when these do not already exist.

However, the most important advantage of using a language like R is that it makes it easy to do data analyses in a way that ensures that they can be exactly repeated. In other words, the biggest advantage of using R, as a language, is not in communicating with the computer, but in communicating to other people what has been done, in a way that is unambiguous. Of course, other people may want to run the same commands in another computer, but still it means that a translation from a set of instructions to the computer into text readable to humans—say the materials and methods section of a paper—and back is avoided together with the ambiguities usually creeping in.

1.2.2 R as a language

R is a computer language designed for data analysis and data visualization, however, in contrast to some other scripting languages, it is, from the point of view of computer programming, a complete language—it is not missing any important feature. In other words, no fundamental operations or data types are lacking (Chambers 2016). I attribute much of its success to the fact that its design achieves a very good balance between simplicity, clarity and generality. R excels at generality thanks to its extensibility at the cost of only a moderate loss of simplicity, while clarity is ensured by enforced documentation of extensions and support for both object-oriented and functional approaches to programming. The same three principles can be also easily respected by user code written in R.

As mentioned above, R started as a free and open-source implementation of the S language (Becker and Chambers 1984; Becker et al. 1988). We will describe the features of the R language in later chapters. Here I mention, for those with programming experience, that it does have some features that make it different from other frequently used programming languages. For example, R does not have the strict type checks of Pascal or C++. It has operators that can take vectors and matrices as operands allowing more concise program statements for such operations than other languages. Writing programs, specially reliable and fast code, requires familiarity with some of these idiosyncracies of the R language. For those using R interactively, or writing short scripts, these idiosyncratic features make life a lot easier by saving typing.

> ☕ Some languages have been standardized, and their grammar has been formally defined. R, in contrast is not standardized, and there is no formal grammar definition. So, the R language is defined by the behavior of the R program.

1.2.3 R as a computer program

The R program itself is open-source, and the source code is available for anybody to inspect, modify and use. A small fraction of users will directly contribute improvements to the R program itself, but it is possible, and those contributions are important in making R reliable. The executable, the R program we actually use, can be built for different operating systems and computer hardware. The members of the R developing team make an important effort to keep the results obtained from calculations done on all the different builds and computer architectures as consistent as possible. The aim is to ensure that computations return consistent results not only across updates to R but also across different operating systems like Linux, Unix (including OS X), and MS-Windows, and computer hardware.

The R program does not have a graphical user interface (GUI), or menus from which to start different types of analyses. Instead, the user types the commands at the R console (Figure 1.1). The same textual commands can also be saved into a text file, line by line, and such a file, called a "script" can substitute repeated typing of the same sequence of commands. When we work at the console typing

```
R R Console                                              ▭ �回 ☒
> print("hello")
[1] "hello"
> |
```

FIGURE 1.1
The R console where the user can type textual commands one by one. Here the user has typed print("Hello") and *entered* it by ending the line of text by pressing the "enter" key. The result of running the command is displayed below the command. The character at the head of the input line, a ">" in this case, is called the command prompt, signaling where a command can be typed in. Commands entered by the user are displayed in red, while results returned by R are displayed in blue.

in commands one by one, we say that we use R interactively. When we run script, we may say that we run a "batch job."

The two approaches described above are part of the R program by itself. However, it is common to use a second program as a front-end or middleman between the user and the R program. Such a program allows more flexibility and has multiple features that make entering commands or writing scripts easier. Computations are still done by exactly the same R program. The simplest option is to use a text editor like Emacs to edit the scripts and then run the scripts in R from within the editor. With some editors like Emacs, rather good integration is possible. However, nowadays there are also Integrated Development Environments (IDEs) available for R. An IDE both gives access to the R console in one window and provides a text editor for writing scripts in another window. Of the available IDEs for R, RStudio is currently the most popular by a wide margin.

1.2.3.1 Using R interactively

A physical terminal (keyboard plus text-only screen) decades ago was how users communicated with computers, and was frequently called a *console*. Nowadays, a text-only interface to a computer, in most cases a window or a pane within a graphical user interface, is still called a console. In our case, the R console (Figure 1.1). This is the native user interface of R.

Typing commands at the R console is useful when one is playing around, rather aimlessly exploring things, or trying to understand how an R function or operator we are not familiar with works. Once we want to keep track of what we are doing, there are better ways of using R, which allow us to keep a record of how an analysis has been carried out. The different ways of using R are not exclusive of each other, so most users will use the R console to test individual commands and plot data during the first stages of exploration. As soon as we decide how we want to plot or analyze the data, it is best to start using scripts. This is not enforced in any way by R, but scripts are what really brings to light the most important advantages of using a programming language for data analysis. In Figure 1.1 we can see how the R console looks. The text in red has been typed in by the user, except for the prompt >, and the text in blue is what R has displayed in response. It is essentially a

```
Console D:/aphalo/Documents/Own_manuscripts/Books/using-r/      =□
> print("Hello")
[1] "Hello"
>
```

FIGURE 1.2
The R console embedded in RStudio. The same commands have been typed in as in Figure 1.1. Commands entered by the user are displayed in purple, while results returned by R are displayed in black.

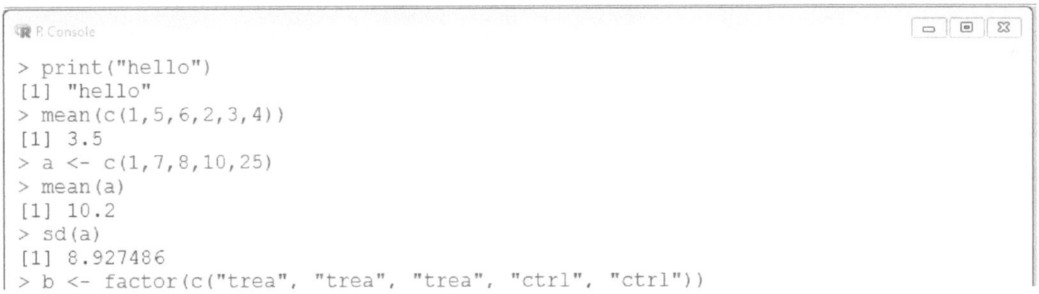

```
R R Console                                      □ ⊡ 83
> print("hello")
[1] "hello"
> mean(c(1,5,6,2,3,4))
[1] 3.5
> a <- c(1,7,8,10,25)
> mean(a)
[1] 10.2
> sd(a)
[1] 8.927486
> b <- factor(c("trea", "trea", "trea", "ctrl", "ctrl"))
```

FIGURE 1.3
The R console after several commands have been entered. Commands entered by the user are displayed in red, while results returned by R are displayed in blue.

dialogue between user and R. The console can *look* different when displayed within an IDE like RStudio, but the only difference is in the appearance of the text rather than in the text itself (cf. Figures 1.1 and 1.2).

The two previous figures showed the result of entering a single command. Figure 1.3 shows how the console looks after the user has entered several commands, each as a separate line of text.

The examples in this book require only the console window for user input. Menu-driven programs are not necessarily bad, they are just unsuitable when there is a need to set very many options and choose from many different actions. They are also difficult to maintain when extensibility is desired, and when independently developed modules of very different characteristics need to be integrated. Textual languages also have the advantage, to be addressed in later chapters, that command sequences can be stored in human- and computer-readable text files. Such files constitute a record of all the steps used, and in most cases, makes it trivial to reproduce the same steps at a later time. Scripts are a very simple and handy way of communicating to other users how to do a given data analysis.

☕ In the console one types commands at the > prompt. When one ends a line by pressing the return or enter key, if the line can be interpreted as an R command, the result will be printed at the console, followed by a new > prompt.

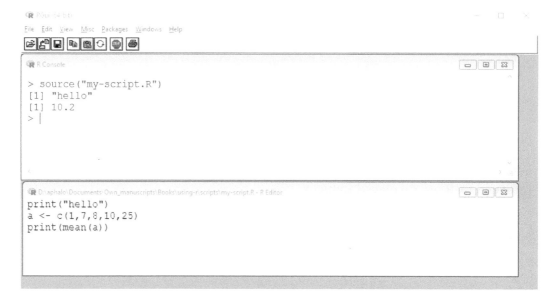

FIGURE 1.4

Screen capture of the R console and editor just after running a script. The upper pane shows the R console, and the lower pane, the script file in an editor.

If the command is incomplete, a + continuation prompt will be shown, and you will be able to type in the rest of the command. For example if the whole calculation that you would like to do is 1 + 2 + 3, if you enter in the console 1 + 2 + in one line, you will get a continuation prompt where you will be able to type 3. However, if you type 1 + 2, the result will be calculated, and printed.

1.2.3.2   Using R in a "batch job"

To run a script we need first to prepare a script in a text editor. Figure 1.4 shows the console immediately after running the script file shown in the text editor. As before, red text, the command `source("my-script.R")`, was typed by the user, and the blue text in the console is what was displayed by R as a result of this action. The title bar of the console, shows "R-console," while the title bar of the editor shows the *path* to the script file that is open and ready to be edited followed by "R-editor."

⚠ When working at the command prompt, most results are printed by default. However, within scripts one needs to use function `print()` explicitly when a result is to be displayed.

A true "batch job" is not run at the R console but at the operating system command prompt, or shell. The shell is the console of the operating system—Linux, Unix, OS X, or MS-Windows. Figure 1.5 shows how running a script at the Windows

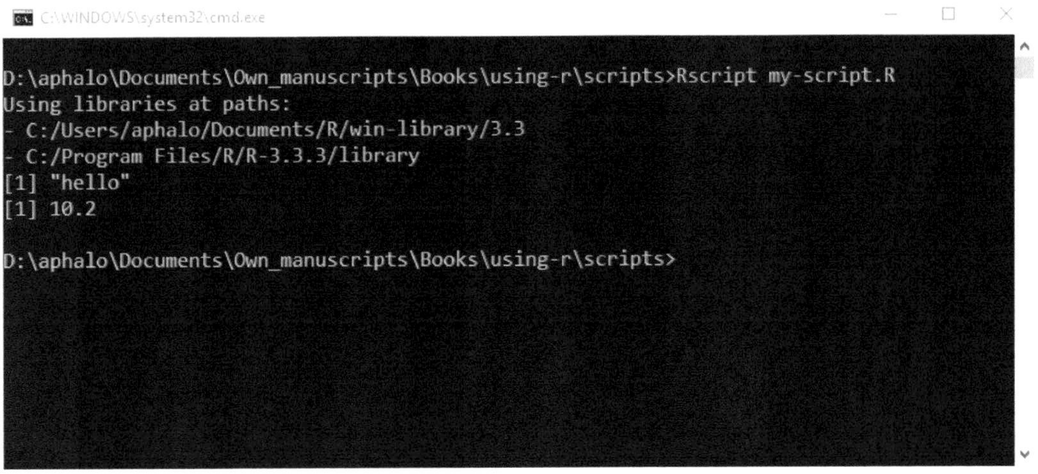

FIGURE 1.5
Screen capture of the MS-Windows command console just after running the same script. Here we use Rscript to run the script; the exact syntax will depend on the operating system in use. In this case, R prints the results at the operating system console or shell, rather than in its own R console.

command prompt looks. A script can be run at the operating system prompt to do time-consuming calculations with the output saved to a file. One may use this approach on a server, say, to leave a large data analysis job running overnight or even for several days.

1.2.3.3   Editors and IDEs

Integrated Development Environments (IDEs) are used when developing computer programs. IDEs provide a centralized user interface from within which the different tools used to create and test a computer program can be accessed and used in coordination. Most IDEs include a dedicated editor capable of syntax highlighting, and even report some mistakes, related to the programming language in use. One could describe such an editor as the equivalent of a word processor with spelling and grammar checking, that can alert about spelling and syntax errors for a computer language like R instead of for a natural language like English. In the case of RStudio, the main, but not only language supported is R. The main window of IDEs usually displays more than one pane simultaneously. From within the RStudio IDE, one has access to the R console, a text editor, a file-system browser, a pane for graphical output, and access to several additional tools such as for installing and updating extension packages. Although RStudio supports very well the development of large scripts and packages, it is currently, in my opinion, also the best possible way of using R at the console as it has the R help system very well integrated both in the editor and R console. Figure 1.6 shows the main window displayed by RStudio after running the same script as shown above at the R console (Figure 1.4) and at the operating system command prompt (Figure 1.5). We can see by comparing these three figures how RStudio is really a layer between the user and an unmodified R executable. The script was sourced by pressing the "Source"

FIGURE 1.6
The RStudio interface just after running the same script. Here we used the "Source" button to run the script. In this case, R prints the results to the R console in the lower left pane.

button at the top of the editor pane. RStudio, in response to this, generated the code needed to source the file and "entered" it at the console, the same console, where we would type any R commands.

When a script is run, if an error is triggered, RStudio automatically finds the location of the error. RStudio supports the concept of projects allowing saving of settings per project. Some features are beyond what you need for everyday data analysis and aimed at package development, such as integration of debugging, traceback on errors, profiling and bench marking of code so as to analyze and improve performance. It integrates support for file version control, which is not only useful for package development, but also for keeping track of the progress or collaboration in the analysis of data.

The version of RStudio that one uses locally, i.e., installed in a computer used locally by a single user, runs with an almost identical user interface on most modern operating systems, such as Linux, Unix, OS X, and MS-Windows. There is also a server version that runs on Linux, and that can be used remotely through a web browser. The user interface is still the same.

RStudio is under active development, and constantly improved. Visit `http://www.rstudio.org/` for an up-to-date description and download and installation instructions. Two books (Hillebrand and Nierhoff 2015; Loo and Jonge 2012) describe and teach how to use RStudio without going in depth into data analysis or statistics, however, as RStudio is under very active development, several recently added important features are not described in these books. You will find tutorials and up-to-date cheat sheets at `http://www.rstudio.org/`.

1.3   Reproducible data analysis

Reproducible data analysis is much more than a fashionable buzzword. Under any situation where accountability is important, from scientific research to decision making in commercial enterprises, industrial quality control and safety and environmental impact assessments, being able to reproduce a data analysis reaching the same conclusions from the same data is crucial. Most approaches to reproducible data analysis are based on automating report generation and including, as part of the report, all the computer commands used to generate the results presented.

A fundamental requirement for reproducibility is a reliable record of what commands have been run on which data. Such a record is especially difficult to keep when issuing commands through menus and dialogue boxes in a graphical user interface or interactively at a console. Even working interactively at the R console using copy and paste to include commands and results in a report is error prone, and laborious.

A further requirement is to be able to match the output of the R commands to the input. If the script saves the output to separate files, then the user will need to take care that the script saved or shared as a record of the data analysis was the one actually used for obtaining the reported results and conclusions. This is another error-prone stage in the reporting of data analysis. To solve this problem an approach was developed, inspired in what is called *literate programming* (Knuth 1984). The idea is that running the script will produce a document that includes the listing of the R code used, the results of running this code and any explanatory text needed to understand and interpret the analysis.

Although a system capable of producing such reports with R, called 'Sweave' (Leisch 2002), has been available for a couple decades, it was rather limited and not supported by an IDE, making its use rather tedious. A more recently developed system called 'knitr' (Xie 2013) together with its integration into RStudio has made the use of this type of reports very easy. The most recent development is what has been called R *notebooks* produced within RStudio. This new feature, can produce the readable report of running the script as an HTML file, displaying the code used interspersed with the results within the viewable file as in earlier approaches. However, this newer approach goes even further: the actual source script used to generate the report is embedded in the HTML file of the report and can be extracted and run very easily and consequently re-used. This means that anyone who gets access to the output of the analysis in human readable form also gets access to the code used to generate the report, in computer executable format.

Because of these recent developments, R is an ideal language to use when the goal of reproducibility is important. During recent years the problem of the lack of reproducibility in scientific research has been broadly discussed and analysed (Gandrud 2015). One of the problems faced when attempting to reproduce experimental work, is reproducing the data analysis. R together with these modern tools can help in avoiding this source of lack of reproducibility.

How powerful are these tools and how flexible? They are powerful and flexible enough to write whole books, such as this very book you are now reading, produced with R, 'knitr' and LaTeX. All pages in the book are generated directly, all figures are generated by R and included automatically, except for the figures in this chapter that have been manually captured from the computer screen. Why am I using this approach? First because I want to make sure that every bit of code as you will see printed, runs without error. In addition, I want to make sure that the output that you will see below every line or chunk of R language code is exactly what R returns. Furthermore, it saves a lot of work for me as author, as I can just update R and all the packages used to their latest version, and build the book again, to keep it up to date and free of errors.

Although the use of these tools is important, they are outside the scope of this book and well described in other books (Gandrud 2015; Xie 2013). Still when writing code, using a consistent style for formatting and indentation, carefully choosing variable names, and adding textual explanations in comments when needed, helps very much with readability for humans. I have tried to be as consistent as possible throughout the whole book in this respect, with only small personal deviations from the usual style.

1.4 Finding additional information

When searching for answers, asking for advice or reading books, you will be confronted with different ways of approaching the same tasks. Do not allow this to overwhelm you; in most cases it will not matter as many computations can be done in R, as in any language, in several different ways, still obtaining the same result. The different approaches may differ mainly in two aspects: 1) how readable to humans are the instructions given to the computer as part of a script or program, and 2) how fast the code runs. Unless computation time is an important bottleneck in your work, just concentrate on writing code that is easy to understand to you and to others, and consequently easy to check and reuse. Of course, do always check any code you write for mistakes, preferably using actual numerical test cases for any complex calculation or even relatively simple scripts. Testing and validation are extremely important steps in data analysis, so get into this habit while reading this book. Testing how every function works, as I will challenge you to do in this book, is at the core of any robust data analysis or computing programming.

⚠ Error messages tend to be terse in R, and may require some lateral thinking and/or "experimentation" to understand the real cause behind problems. When you are not sure you understand how some command works, it is useful in many cases to try simple examples for which you know the correct answer and see if you can reproduce them with R. Because of this, this book includes some code examples that trigger errors. Learning to interpret error messages is part of what is needed to become a proficient user of R. To test your understanding of how a code statement or function works, it is good to try your hand at testing its limits, testing which variations of a piece code are valid or not.

1.4.1   R's built-in help

To access help pages through the command prompt we use function `help()` or a question mark. Every object exported by an R package (functions, methods, classes, data) is documented. Sometimes a single help page documents several R objects. Usually at the end of the help pages, some examples are given, which tend to help very much in learning how to use the functions described. For example, one can search for a help page at the R console.

```
help("sum")
?sum
```

> Look at help for some other functions like `mean()`, `var()`, `plot()` and, why not, `help()` itself!
>
> ```
> help(help)
> ```

When using RStudio there are easier ways of navigating to a help page than using function `help()`, for example, with the cursor on the name of a function in the editor or console, pressing the F1 key opens the corresponding help page in the help pane. Letting the cursor hover for a few seconds over the name of a function at the R console will open "bubble help" for it. If the function is defined in a script or another file that is open in the editor pane, one can directly navigate from the line where the function is called to where it is defined. In RStudio one can also search for help through the graphical interface.

In addition to help pages, R's distribution includes useful manuals as PDF or HTML files. These can be accessed most easily through the Help menu in RStudio or RGUI. Extension packages provide help pages for the functions and data they export. When a package is loaded into an R session, its help pages are added to the native help of R. In addition to these individual help pages, each package provides an index of its corresponding help pages for users to browse. Many packages, contain *vignettes* such as User Guides or articles describing the algorithms used.

There are some web sites that give access to R documentation through a web server. These sites can be very convenient when exploring whether a certain package could be useful for a certain problem, as they allow browsing and searching the documentation without need of installing the packages. Some package maintainers have web sites with additional documentation for their own packages. The DESCRIPTION or README of packages provide contact information for the maintainer, links to web sites, and instructions on how to report bugs. As packages are contributed by independent authors, they should be cited in addition to citing R itself. R function `citation()` when called with the name of a package as its argument provides the reference that should be cited for the package, and without an explicit argument, the reference to cite for the version of R in use.

```
citation()
##
## To cite R in publications use:
##
##    R Core Team (2020). R: A language and environment for statistical
##    computing. R Foundation for Statistical Computing, Vienna, Austria.
##    URL https://www.R-project.org/.
##
## A BibTeX entry for LaTeX users is
##
##    @Manual{,
##      title = {R: A Language and Environment for Statistical Computing},
##      author = {{R Core Team}},
##      organization = {R Foundation for Statistical Computing},
##      address = {Vienna, Austria},
##      year = {2020},
##      url = {https://www.R-project.org/},
##    }
##
## We have invested a lot of time and effort in creating R, please cite it
## when using it for data analysis. See also 'citation("pkgname")' for
## citing R packages.
```

> Look at the help page for function `citation()` for a discussion of why it is important for users to cite R and packages when using them.

1.4.2 Obtaining help from online forums

When consulting help pages, vignettes, and possibly books at hand fails to provide the information needed, the next step to follow is to search internet forums for existing answers to one's questions. When these steps fail to solve a problem, then it is time to ask for help, either from local experts or by posting your own question in a suitable online forum. When posting requests for help, one needs to abide by what is usually described as "netiquette."

1.4.2.1 Netiquette

In most internet forums, a certain behavior is expected from those asking and answering questions. Some types of misbehavior, like use of offensive or inappropriate language, will usually result in the user losing writing rights in a forum. Occasional minor misbehavior, will usually result in the original question not being answered and instead the problem highlighted in the reply. In general following the steps listed below will greatly increase your chances of getting a detailed and useful answer.

• Do your homework: first search for existing answers to your question, both online and in the documentation. (Do mention that you attempted this without success when you post your question.)

- Provide a clear explanation of the problem, and all the relevant information. Say if it concerns R, the version, operating system, and any packages loaded and their versions.

- If at all possible, provide a simplified and short, but self-contained, code example that reproduces the problem (sometimes called *reprex*).

- Be polite.

- Contribute to the forum by answering other users' questions when you know the answer.

1.4.2.2　StackOverflow

Nowadays, StackOverflow (`http://stackoverflow.com/`) is the best question-and-answer (Q&A) support site for R. In most cases, searching for existing questions and their answers, will be all that you need to do. If asking a question, make sure that it is really a new question. If there is some question that looks similar, make clear how your question is different.

StackOverflow has a user-rights system based on reputation, and questions and answers can be up- and down-voted. Those with the most up-votes are listed at the top of searches. If the questions or answers you write are up-voted, after you accumulate enough reputation, you acquire badges and rights, such as editing other users' questions and answers or later on, even deleting wrong answers or off-topic questions from the system. This sounds complicated, but works extremely well at ensuring that the base of questions and answers is relevant and correct, without relying on nominated *moderators*. When using StackOverflow, do contribute by accepting correct answers, up-voting questions and answers that you find useful, down-voting those you consider poor, and flagging or correcting errors you may discover.

1.4.2.3　Reporting bugs

Being careful in the preparation of a reproducible example is especially important when you intend to report a bug to the maintainer of any piece of software. For the problem to be fixed, the person revising the code, needs to be able to reproduce the problem, and after modifying the code, needs to be able to test if the problem has been solved or not. However, even if you are facing a problem caused by your misunderstanding of how R works, the simpler the example, the more likely that someone will quickly realize what your intention was when writing the code that produces a result different from what you expected.

How to prepare a reproducible example ("reprex"). A *reprex* is a self-contained and as simple as possible piece of computer code that triggers (and so demonstrates) a problem. If possible, when you need to use data, either use a data set included in base R or generate artificial data within the reprex code. If you can reproduce the problem only with your own data, then you need to provide a minimal subset of it that triggers the problem.

While preparing the *reprex* you will need to simplify the code, and sometimes this step allows you to diagnose the problem. Always, before posting a reprex online, it is wise to check it with the latest versions of R and any package being used.

I would say that about two out of three times I prepare a *reprex*, it allows me to much better understand the problem and find the root of the problem and a solution or a work-around on my own.

1.5   What is needed to run the examples in this book?

The book is written with the expectation that you will run most of the code examples and try as many other variations as needed until you are sure you understand the basic "rules" of the R language and how each function or command described works. As mentioned above, you are expected to use this book as a travel guide for your exploration of the world of R.

R is all that is needed to work through all the examples in this book, but it is not a convenient way of doing this. I recommend that you use an editor or an IDE, in particular RStudio . RStudio is user friendly, actively maintained, free, open-source and available both in desktop and server versions. The desktop version runs on MS-Windows, Linux, and OS X and other Unix distributions.

Of course when choosing which editor to use, personal preferences and previous familiarity play an important role. Currently, for the development of packages, I use RStudio exclusively. For writing this book I have used both RStudio and the text editor WinEdt which has support for R together with excellent support for LaTeX. When working on a large project or collaborating with other data analysts or researchers, one big advantage of a system based on plain text files such as R scripts, is that the same files can be edited with different programs and under different operating systems as needed or wished by the different persons involved in a project.

When I started using R, nearly two decades ago, I was using other editors, using the operating system shell a lot more, and struggling with debugging as no IDE was available. The only reasonably good integration with an editor was for Emacs, which was widely available only under Unix-like systems. Given my past experience, I encourage you to use an IDE for R. RStudio is nowadays very popular, but if you do not like it, need a different set of features, such as integration with ImageJ, or are already familiar with the Eclipse IDE, you may want to try the Bio7 IDE, available from http://bio7.org.

The examples in this book make use of several freely available R extension packages, which can be installed from CRAN. One of them, 'learnrbook', also available through CRAN, contains data sets and files specific to this book. The 'learnrbook' package contains installation instructions and saved lists of the names of all other packages used in the book. Instructions on installing R, Git, RStudio, compilers and other tools are available online. In many cases the IT staff at your employer or school will know how to install them, or they may even be included in the default computer setup. In addition, a web site supporting the book will be available at: `http://www.learnr-book.info`.

1.6 Further reading

Suggestions for further reading are dependent on how you plan to use R. If you envision yourself running batch jobs under Linux or Unix, you would profit from learning to write shell scripts. Because `bash` is widely used nowadays, *Learning the bash Shell* (Newham and Rosenblatt 2005) can be recommended. If you aim at writing R code that is going to be reused, and have some familiarity with C, C++ or Java, reading *The Practice of Programming* (Kernighan and Pike 1999) will provide a mostly language-independent view of programming as an activity and help you master the all-important tricks of the trade.

2

The R language: "Words" and "sentences"

The desire to economize time and mental effort in arithmetical computations, and to eliminate human liability to error, is probably as old as the science of arithmetic itself.

Howard Aiken
Proposed automatic calculating machine, 1937; reprinted 1964

2.1 Aims of this chapter

In my experience, for those not familiar with computer programming languages, the best first step in learning the R language is to use it interactively by typing textual commands at the *console* or command line. This will teach not only the syntax and grammar rules, but also give you a glimpse at the advantages and flexibility of this approach to data analysis.

In the first part of the chapter we will use R to do everyday calculations that should be so easy and familiar that you will not need to think about the operations themselves. This easy start will give you a chance to focus on learning how to issue textual commands at the command prompt.

Later in the chapter, you will gradually need to focus more on the R language and its grammar and less on how commands are entered. By the end of the chapter you will be familiar with most of the kinds of "words" used in the R language and you will be able to write simple "sentences" in R.

Along the chapter, I will occasionally show the equivalent of the R code in mathematical notation. If you are not familiar with the mathematical notation, you can safely ignore it, as long as you understand the R code.

2.2   Natural and computer languages

Computer languages have strict rules and interpreters and compilers are unforgiving about errors. They will issue error messages, but in contrast to human readers or listeners, will not guess your intentions and continue. However, computer languages have a much smaller set of words than natural languages, such as English. If you are new to computer programming, understanding the parallels between computer and natural languages may be useful.

One can think of constant values and variables (values stored under a name) as nouns and of operators and functions as verbs. A complete command, or statement, is the equivalent of a natural language sentence: "a comprehensible utterance." The simple statement a + 1 has three components: a, a variable, +, an operator and 1 a constant. The statement sqrt(4) has two components, a function sqrt() and a numerical constant 4. We say that "to compute $\sqrt{4}$ we *call* sqrt() with 4 as its *argument*."

In later chapters you will learn how to write compound statements, the equivalent of natural-language paragraphs, and scripts, the equivalent of essays. You will also learn how to define new verbs, user-defined functions and operators, and new nouns, user-defined classes.

2.3   Numeric values and arithmetic

When working in R with arithmetic expressions, the normal mathematical precedence rules are respected, but parentheses can be used to alter this order. Parentheses can be nested, but in contrast to the usual practice in mathematics, the same parenthesis symbol is used at all nesting levels.

> ⌨ Both in mathematics and programming languages *operator precedence rules* determine which subexpressions are evaluated first and which later. Contrary to primitive electronic calculators, R evaluates numeric expressions containing operators according to the rules of mathematics. In the expression $3 + 2 \times 3$, the product 2×3 has precedence over the addition, and is evaluated first, yielding as the result of the whole expression, 9. In programming languages, similar rules apply to all operators, even those taking as operands non-numeric values.

It is important to keep in mind that in R trigonometric functions interpret numeric values representing angles as being expressed in radians.

The equivalent of the math expression

$$\frac{3 + e^2}{\sin \pi}$$

is, in R, written as follows:

```
(3 + exp(2)) / sin(pi)
## [1] 8.483588e+16
```

It can be seen above that mathematical constants and functions are part of the R language. One thing to remember when translating complex fractions as above into R code, is that in arithmetic expressions the bar of the fraction generates a grouping that alters the normal precedence of operations. In contrast, in an R expression this grouping must be explicitly signaled with additional parentheses.

If you are in doubt about how precedence rules work, you can add parentheses to make sure the order of computations is the one you intend. Redundant parentheses have no effect.

```
1 + 2 * 3
## [1] 7
```

```
1 + (2 * 3)
## [1] 7
```

```
(1 + 2) * 3
## [1] 9
```

The number of opening (left side) and closing (right side) parentheses must be balanced, and they must be located so that each enclosed term is a valid mathematical expression. For example, while (1 + 2) * 3 is valid, (1 +) 2 * 3 is a syntax error as 1 + is incomplete and cannot be calculated.

> Here results are not shown. These are examples for you to type at the command prompt. In general you should not skip them, as in many cases, as with the statements highlighted with comments in the code chunk below, they have something to teach or demonstrate. You are strongly encouraged to *play*, in other words, create new variations of the examples and execute them to explore how R works.
>
> ```
> 1 + 1
> 2 * 2
> 2 + 10 / 5
> (2 + 10) / 5
> 10^2 + 1
> sqrt(9)
> pi # whole precision not shown when printing
> print(pi, digits = 22)
> sin(pi) # oops! Read on for explanation.
> log(100)
> log10(100)
> log2(8)
> exp(1)
> ```

Variables are used to store values. After we *assign* a value to a variable, we can use the name of the variable in place of the stored value. The "usual" assignment operator is <-. In R, all names, including variable names, are case sensitive.

Variables a and A are two different variables. Variable names can be long in R although it is not a good idea to use very long names. Here I am using very short
names, something that is usually also a very bad idea. However, in the examples
in this chapter where the stored values have no connection to the real world, simple names emphasize their abstract nature. In the chunk below, a and b are ar-
bitrarily chosen variable names; I could have used names like `my.variable.a` or
`outside.temperature` if they had been useful to convey information.

```
a <- 1
a + 1
## [1] 2

a
## [1] 1

b <- 10
b <- a + b
b
## [1] 11

3e-2 * 2.0
## [1] 0.06
```

Entering the name of a variable *at the R console* implicitly calls function `print()`
displaying the stored value on the console. The same applies to any other statement entered *at the R console*: `print()` is implicitly called with the result of executing the statement as its argument.

```
a
## [1] 1

print(a)
## [1] 1

a + 1
## [1] 2

print(a + 1)
## [1] 2
```

There are some syntactically legal statements that are not very frequently
used, but you should be aware that they are valid, as they will not trigger error
messages, and may surprise you. The most important thing is to write code
consistently. The "backwards" assignment operator -> and resulting code like
1 -> a are valid but less frequently used. The use of the equals sign (=) for
assignment in place of <- although valid is discouraged. Chaining assignments
as in the first statement below can be used to signal to the human reader that
a, b and c are being assigned the same value.

```
a <- b <- c <- 0.0
a
b
c
1 -> a
a
a = 3
a
```

In R, all numbers belong to mode numeric (we will discuss the concepts of *mode* and *class* in section 2.8 on page 41). We can query if the mode of an object is numeric with function is.numeric().

```
mode(1)
## [1] "numeric"

a <- 1
is.numeric(a)
## [1] TRUE
```

Because numbers can be stored in different formats, requiring different amounts of computer memory per value, most computing languages implement several different types of numbers. In most cases R's numeric() can be used everywhere that a number is expected. However, in some cases it has advantages to explicitly indicate that we will store or operate on whole numbers, in which case we can use class integer, with integer constants indicated by a trailing capital "L," as in 32L.

```
is.numeric(1L)
## [1] TRUE

is.integer(1L)
## [1] TRUE

is.double(1L)
## [1] FALSE
```

Real numbers are a mathematical abstraction, and do not have an exact equivalent in computers. Instead of Real numbers, computers store and operate on numbers that are restricted to a broad but finite range of values and have a finite resolution. They are called, *floats* (or *floating-point* numbers); in R they go by the name of double and can be created with the constructor double().

```
is.numeric(1)
## [1] TRUE

is.integer(1)
## [1] FALSE

is.double(1)
## [1] TRUE
```

The name `double` originates from the C language, in which there are different types of floats available. With the name `double` used to mean "double-precision floating-point numbers." Similarly, the use of `L` stems from the `long` type in C, meaning "long integer numbers."

Numeric variables can contain more than one value. Even single numbers are in R `vector`s of length one. We will later see why this is important. As you have seen above, the results of calculations were printed preceded with [1]. This is the index or position in the vector of the first number (or other value) displayed at the head of the current line.

One can use `c()` "concatenate" to create a vector from other vectors, including vectors of length 1, such as the `numeric` constants in the statements below.

```
a <- c(3, 1, 2)
a
## [1] 3 1 2

b <- c(4, 5, 0)
b
## [1] 4 5 0

c <- c(a, b)
c
## [1] 3 1 2 4 5 0

d <- c(b, a)
d
## [1] 4 5 0 3 1 2
```

Method `c()` accepts as arguments two or more vectors and concatenates them, one after another. Quite frequently we may need to insert one vector in the middle of another. For this operation, `c()` is not useful by itself. One could use indexing combined with `c()`, but this is not needed as R provides a function capable of directly doing this operation. Although it can be used to "insert" values, it is named `append()`, and by default, it indeed appends one vector at the end of another.

```
append(a, b)
## [1] 3 1 2 4 5 0
```

The output above is the same as for `c(a, b)`, however, `append()` accepts as an argument an index position after which to "append" its second argument. This results in an *insert* operation when the index points at any position different from the end of the vector.

```
append(a, values = b, after = 2L)
## [1] 3 1 4 5 0 2
```

Both `c()` and `append()` can also be used with lists.

> One can create sequences using function `seq()` or the operator `:`, or repeat values using function `rep()`. In this case, I leave to the reader to work out the rules by running these and his/her own examples, with the help of the documentation, available through `help(seq)` and `help(rep)`.
>
> ```
> a <- -1:5
> a
> b <- 5:-1
> b
> c <- seq(from = -1, to = 1, by = 0.1)
> c
> d <- rep(-5, 4)
> d
> ```

Next, something that makes R different from most other programming languages: vectorized arithmetic. Operators and functions that are vectorized accept, as arguments, vectors of arbitrary length, in which case the result returned is equivalent to having applied the same function or operator individually to each element of the vector.

```
a + 1 # we add one to vector a defined above
## [1] 4 2 3

(a + 1) * 2
## [1] 8 4 6

a + b
## [1] 7 6 2

a - a
## [1] 0 0 0
```

As it can be seen in the first line above, another peculiarity of R, is what is frequently called "recycling" of arguments: as vector a is of length 6, but the constant 1 is a vector of length 1, this short constant vector is extended, by recycling its value, into a vector of six ones—i.e., a vector of the same length as the longest vector in the statement, a.

Make sure you understand what calculations are taking place in the chunk above, and also the one below.

```
a <- rep(1, 6)
a
## [1] 1 1 1 1 1 1

a + 1:2
```

```
## [1] 2 3 2 3 2 3

a + 1:3
## [1] 2 3 4 2 3 4

a + 1:4

## Warning in a + 1:4: longer object length is not a multiple of shorter object
length
## [1] 2 3 4 5 2 3
```

☕ A useful thing to know: a vector can have length zero. Vectors of length zero may seem at first sight quite useless, but in fact they are very useful. They allow the handling of "no input" or "nothing to do" cases as normal cases, which in the absence of vectors of length zero would require to be treated as special cases. I describe here a useful function, `length()` which returns the length of a vector or list.

```
z <- numeric(0)
z
## numeric(0)

length(z)
## [1] 0
```

Vectors and lists of length zero, behave in most cases, as expected—e.g., they can be concatenated as shown here.

```
length(c(a, numeric(0), b))
## [1] 9

length(c(a, b))
## [1] 9
```

Many functions, such as R's maths functions and operators, will accept numeric vectors of length zero as valid input, returning also a vector of length zero, issuing neither a warning nor an error message. In other words, *these are valid operations* in R.

```
log(numeric(0))
## numeric(0)

5 + numeric(0)
## numeric(0)
```

Even when of length zero, vectors do have to belong to a class acceptable for the operation.

It is possible to *remove* variables from the workspace with `rm()`. Function `ls()` returns a *list* of all objects visible in the current environment, or by supplying a `pattern` argument, only the objects with names matching the `pattern`. The pattern is given as a regular expression, with [] enclosing alternative matching characters, ∧ and $, indicating the extremes of the name (start and end, respectively). For example, `"∧z$"` matches only the single character 'z' while `"∧z"` matches any name starting with 'z'. In contrast `"∧[zy]$"` matches both 'z' and 'y' but neither 'zy' nor 'yz', and `"∧[a-z]"` matches any name starting with a lowercase ASCII letter. If you are using RStudio, all objects are listed in the Environment pane, and the search box of the panel can be used to find a given object.

```
ls(pattern="∧z$")
## [1] "z"

rm(z)
ls(pattern="∧z$")
## character(0)
```

There are some special values available for numbers. NA meaning "not available" is used for missing values. Calculations can also yield the following values NaN "not a number", Inf and -Inf for ∞ and −∞. As you will see below, calculations yielding these values do **not** trigger errors or warnings, as they are arithmetically valid. Inf and -Inf are also valid numerical values for input and constants.

```
a <- NA
a
## [1] NA

-1 / 0
## [1] -Inf

1 / 0
## [1] Inf

Inf / Inf
## [1] NaN

Inf + 4
## [1] Inf

b <- -Inf
b * -1
## [1] Inf
```

Not available (NA) values are very important in the analysis of experimental data, as frequently some observations are missing from an otherwise complete data set due to "accidents" during the course of an experiment. It is important to understand how to interpret NA's. They are simple placeholders for something that is unavailable, in other words, *unknown*.

```
A <- NA
A
## [1] NA

A + 1
## [1] NA

A + Inf
## [1] NA
```

> 📊 **When to use vectors of length zero, and when NAS?** Make sure you understand the logic behind the different behavior of functions and operators with respect to NA and numeric() or its equivalent numeric(0). What do they represent? Why NA s are not ignored, while vectors of length zero are?
>
> ```
> 123 + numeric()
> 123 + NA
> ```
>
> *Model answer:* NA is used to signal a value that "was lost" or "was expected" but is unavailable because of some accident. A vector of length zero, represents no values, but within the normal expectations. In particular, if vectors are expected to have a certain length, or if index positions along a vector are meaningful, then using NA is a must.

Any operation, even tests of equality, involving one or more NA's return an NA. In other words, when one input to a calculation is unknown, the result of the calculation is unknown. This means that a special function is needed for testing for the presence of NA values.

```
is.na(c(NA, 1))
## [1]  TRUE FALSE
```

In the example above, we can also see that `is.na()` is vectorized, and that it applies the test to each of the two elements of the vector individually, returning the result as a logical vector of length two.

One thing to be aware of are the consequences of the fact that numbers in computers are almost always stored with finite precision and/or range: the expectations derived from the mathematical definition of Real numbers are not always fulfilled. See the box on page 33 for an in-depth explanation.

```
1 - 1e-20
## [1] 1
```

When comparing `integer` values these problems do not exist, as integer arithmetic is not affected by loss of precision in calculations restricted to integers. Because of the way integers are stored in the memory of computers, within the representable range, they are stored exactly. One can think of computer integers as a subset of whole numbers restricted to a certain range of values.

```
1L + 3L
## [1] 4

1L * 3L
## [1] 3

1L %/% 3L
## [1] 0

1L %% 3L
## [1] 1

1L / 3L
## [1] 0.3333333
```

The last statement in the example immediately above, using the "usual" division operator yields a floating-point `double` result, while the integer division operator `%/%` yields an `integer` result, and `%%` returns the remainder from the integer division. If as a result of an operation the result falls outside the range of representable values, the returned value is NA.

```
1000000L * 1000000L
```

```
## Warning in 1000000L * 1000000L: NAs produced by integer overflow
## [1] NA
```

Both doubles and integers are considered numeric. In most situations, conversion is automatic and we do not need to worry about the differences between these two types of numeric values. The next chunk shows returned values that are either TRUE or FALSE. These are `logical` values that will be discussed in the next section.

```
is.numeric(1L)
## [1] TRUE
```

```
is.integer(1L)
## [1] TRUE
```

```
is.double(1L)
## [1] FALSE
```

```
is.double(1L / 3L)
## [1] TRUE
```

```
is.numeric(1L / 3L)
## [1] TRUE
```

> Study the variations of the previous example shown below, and explain why the two statements return different values. Hint: 1 is a `double` constant. You can use `is.integer()` and `is.double()` in your explorations.
>
> ```
> 1 * 1000000L * 1000000L
> 1000000L * 1000000L * 1
> ```

Both when displaying numbers or as part of computations, we may want to decrease the number of significant digits or the number of digits after the decimal marker. Be aware that in the examples below, even if printing is being done by default, these functions return `numeric` values that are different from their input and can be stored and used in computations. Function `round()` is used to round numbers to a certain number of decimal places after or before the decimal marker, while `signif()` rounds to the requested number of significant digits.

```
round(0.0124567, digits = 3)
## [1] 0.012
```

```
signif(0.0124567, digits = 3)
```

```
## [1] 0.0125
```

```
round(1789.1234, digits = 3)
## [1] 1789.123
```

```
signif(1789.1234, digits = 3)
## [1] 1790
```

```
round(1789.1234, digits = -1)
## [1] 1790
```

```
a <- 0.12345
b <- round(a, digits = 2)
a == b
## [1] FALSE
```

```
a - b
## [1] 0.00345
```

```
b
## [1] 0.12
```

Being `digits`, the second parameter of these functions, the argument can also be passed by position. However, code is usually easier to understand for humans when parameter names are made explicit.

```
round(0.0124567, digits = 3)
## [1] 0.012
```

```
round(0.0124567, 3)
## [1] 0.012
```

Functions `trunc()` and `ceiling()` return the non-fractional part of a numeric value as a new numeric value. They differ in how they handle negative values, and neither of them rounds the returned value to the nearest whole number.

What does value truncation mean? Function `trunc()` truncates a numeric value, but it does not return an `integer`.

- Explore how `trunc()` and `ceiling()` differ. Test them both with positive and negative values.

- **Advanced** Use function `abs()` and operators + and − to reproduce the output of `trunc()` and `ceiling()` for the different inputs.

- Can `trunc()` and `ceiling()` be considered type conversion functions in R?

2.4 Logical values and Boolean algebra

What in Mathematics are usually called Boolean values, are called `logical` values in R. They can have only two values TRUE and FALSE, in addition to NA (not available). They are vectors as all other atomic types in R (by *atomic* we mean that each value is not composed of "parts"). There are also logical operators that allow Boolean algebra. In the chunk below we operate on `logical` vectors of length one.

```r
a <- TRUE
b <- FALSE
mode(a)
## [1] "logical"

a
## [1] TRUE

!a # negation
## [1] FALSE

a && b # logical AND
## [1] FALSE

a || b # logical OR
## [1] TRUE

xor(a, b) # exclusive OR
## [1] TRUE
```

As with arithmetic operators, vectorization is available with *some* logical operators. The availability of two kinds of logical operators is one of the most troublesome aspects of the R language for beginners. Pairs of "equivalent" logical operators behave differently, use similar syntax and use similar symbols! The vectorized operators have single-character names & and |, while the non-vectorized ones have double-character names && and ||. There is only one version of the negation operator ! that is vectorized. In some, but not all cases, a warning will indicate that there is a possible problem.

```r
a <- c(TRUE,FALSE)
b <- c(TRUE,TRUE)
a
## [1]  TRUE FALSE

b
## [1] TRUE TRUE

a & b # vectorized AND
## [1]  TRUE FALSE

a | b # vectorized OR
## [1] TRUE TRUE

a && b # not vectorized
## [1] TRUE

a || b # not vectorized
## [1] TRUE
```

Functions any() and all() take zero or more logical vectors as their arguments, and return a single logical value "summarizing" the logical values in the vectors. Function all() returns TRUE only if all values in the vectors passed as arguments are TRUE, and any() returns TRUE unless all values in the vectors are FALSE.

```
any(a)
## [1] TRUE

all(a)
## [1] FALSE

any(a & b)
## [1] TRUE

all(a & b)
## [1] FALSE
```

Another important thing to know about logical operators is that they "short-cut" evaluation. If the result is known from the first part of the statement, the rest of the statement is not evaluated. Try to understand what happens when you enter the following commands. Short-cut evaluation is useful, as the first condition can be used as a guard protecting a later condition from being evaluated when it would trigger an error.

```
TRUE || NA
## [1] TRUE

FALSE || NA
## [1] NA

TRUE && NA
## [1] NA

FALSE && NA
## [1] FALSE

TRUE && FALSE && NA
## [1] FALSE

TRUE && TRUE && NA
## [1] NA
```

When using the vectorized operators on vectors of length greater than one, 'short-cut' evaluation still applies for the result obtained at each index position.

```
a & b & NA
## [1]    NA FALSE

a & b & c(NA, NA)
## [1]    NA FALSE

a | b | c(NA, NA)
## [1] TRUE TRUE
```

> Based on the description of "recycling" presented on page 23 for numeric operators, explore how "recycling" works with vectorized logical operators. Create logical vectors of different lengths (including length one) and *play* by writing several code statements with operations on them. To get you started, one example is given below. Execute this example, and then create and run your own, making sure that you understand why the values returned are what they are. Sometimes, you will need to devise several examples or test cases to tease out of R an understanding of how a certain feature of the language works, so do not give up early, and make use of your imagination!

```
x <- c(TRUE, FALSE, TRUE, NA)
x & FALSE
x | c(TRUE, FALSE)
```

2.5 Comparison operators and operations

Comparison operators return vectors of `logical` values as results.

```
1.2 > 1.0
## [1] TRUE

1.2 >= 1.0
## [1] TRUE

1.2 == 1.0 # be aware that here we use two = symbols
## [1] FALSE

1.2 != 1.0
## [1] TRUE

1.2 <= 1.0
## [1] FALSE

1.2 < 1.0
## [1] FALSE

a <- 20
a < 100 && a > 10
## [1] TRUE
```

These operators can be used on vectors of any length, returning as a result a logical vector as long as the longest operand. In other words, they behave in the same way as the arithmetic operators described on page 23: their arguments are recycled when needed. Hint: if you do not know what to expect as a value for the vector returned by `1:10`, execute the statement `print(a)` after the first code statement below, or, alternatively, `1:10` without saving the result to a variable.

```
a <- 1:10
a > 5
## [1] FALSE FALSE FALSE FALSE FALSE  TRUE  TRUE  TRUE  TRUE  TRUE

a < 5
## [1]  TRUE  TRUE  TRUE  TRUE FALSE FALSE FALSE FALSE FALSE FALSE

a == 5
## [1] FALSE FALSE FALSE FALSE  TRUE FALSE FALSE FALSE FALSE FALSE

all(a > 5)
## [1] FALSE

any(a > 5)
## [1] TRUE

b <- a > 5
b
## [1] FALSE FALSE FALSE FALSE FALSE  TRUE  TRUE  TRUE  TRUE  TRUE

any(b)
## [1] TRUE

all(b)
## [1] FALSE
```

Precedence rules also apply to comparison operators and they can be overridden by means of parentheses.

```
a > 2 + 3
## [1] FALSE FALSE FALSE FALSE FALSE  TRUE  TRUE  TRUE  TRUE  TRUE

(a > 2) + 3
## [1] 3 3 4 4 4 4 4 4 4 4
```

> Use the statement below as a starting point in exploring how precedence works when logical and arithmetic operators are part of the same statement. *Play* with the example by adding parentheses at different positions and based on the returned values, work out the default order of operator precedence used for the evaluation of the example given below.
>
> ```
> a <- 1:10
> a > 3 | a + 2 < 3
> ```

Again, be aware of "short-cut evaluation". If the result does not depend on the missing value, then the result, TRUE or FALSE is returned. If the presence of the NA makes the end result unknown, then NA is returned.

```
c <- c(a, NA)
c > 5
## [1] FALSE FALSE FALSE FALSE FALSE  TRUE  TRUE  TRUE  TRUE  TRUE    NA
```

```
all(c > 5)
## [1] FALSE
```

```
any(c > 5)
## [1] TRUE
```

```
all(c < 20)
## [1] NA
```

```
any(c > 20)
## [1] NA
```

```
is.na(a)
##   [1] FALSE FALSE FALSE FALSE FALSE FALSE FALSE FALSE FALSE FALSE
```

```
is.na(c)
##   [1] FALSE FALSE FALSE FALSE FALSE FALSE FALSE FALSE FALSE FALSE  TRUE
```

```
any(is.na(c))
## [1] TRUE
```

```
all(is.na(c))
## [1] FALSE
```

The behavior of many of base-R's functions when NAs are present in their input arguments can be modified. TRUE passed as an argument to parameter na.rm, results in NA values being *removed* from the input **before** the function is applied.

```
all(c < 20)
## [1] NA
```

```
any(c > 20)
## [1] NA
```

```
all(c < 20, na.rm=TRUE)
## [1] TRUE
```

```
any(c > 20, na.rm=TRUE)
## [1] FALSE
```

📟   Here I give some examples for which the finite resolution of computer machine floats, as compared to Real numbers as defined in mathematics, can cause serious problems. In R, numbers that are not integers are stored as *double-precision floats*. In addition to having limits to the largest and smallest numbers that can be represented, the precision of floats is limited by the number of significant digits that can be stored. Precision is usually described by "epsilon" (ϵ), abbreviated *eps*, defined as the largest value of ϵ for which $1 + \epsilon = 1$. The finite resolution of floats can lead to unexpected results when testing for equality. In the second example below, the result of the subtraction is still exactly 1 due to insufficient resolution.

```
0 - 1e-20
## [1] -1e-20

1 - 1e-20
## [1] 1
```

The finiteness of floats also affects tests of equality, which is more likely to result in errors with important consequences.

```
1e20 == 1 + 1e20
## [1] TRUE

1 == 1 + 1e-20
## [1] TRUE

0 == 1e-20
## [1] FALSE
```

As R can run on different types of computer hardware, the actual machine limits for storing numbers in memory may vary depending on the type of processor and even compiler used to build the R program executable. However, it is possible to obtain these values at run time from the variable .Machine, which is part of the R language. Please see the help page for .Machine for a detailed and up-to-date description of the available constants.

```
.Machine$double.eps
## [1] 2.220446e-16

.Machine$double.neg.eps
## [1] 1.110223e-16

.Machine$double.max
## [1] 1024

.Machine$double.min
## [1] -1022
```

The last two values refer to the exponents of 10, rather than the maximum and minimum size of numbers that can be handled as objects of class double. Values outside these limits are stored as -Inf or Inf and enter arithmetic as infinite values according the mathematical rules.

```
1e1026
## [1] Inf

1e-1026
## [1] 0

Inf + 1
## [1] Inf

-Inf + 1
## [1] -Inf
```

As `integer` values are stored in machine memory without loss of precision, epsilon is not defined for `integer` values.

```
.Machine$integer.max
## [1] 2147483647

2147483699L
## [1] 2147483699
```

In those statements in the chunk below where at least one operand is `double` the `integer` operands are *promoted* to `double` before computation. A similar promotion does not take place when operations are among `integer` values, resulting in *overflow*, meaning numbers that are too big to be represented as `integer` values.

```
2147483600L + 99L

## Warning in 2147483600L + 99L: NAs produced by integer overflow
## [1] NA

2147483600L + 99
## [1] 2147483699

2147483600L * 2147483600L

## Warning in 2147483600L * 2147483600L: NAs produced by integer overflow
## [1] NA

2147483600L * 2147483600
## [1] 4.611686e+18
```

We see next that the exponentiation operator ∧ forces the promotion of its arguments to `double`, resulting in no overflow. In contrast, as seen above, the multiplication operator * operates on integers resulting in overflow.

```
2147483600L * 2147483600L

## Warning in 2147483600L * 2147483600L: NAs produced by integer overflow
## [1] NA

2147483600L^2L
## [1] 4.611686e+18
```

⚠ In many situations, when writing programs one should avoid testing for equality of floating point numbers ('floats'). Here we show how to gracefully handle rounding errors. As the example shows, rounding errors may accumulate, and in practice `.Machine$double.eps` is not always a good value to safely use in tests for "zero," and a larger value may be needed. Whenever possible according to the logic of the calculations, it is best to test for inequalities, for example using x <= 1.0 instead of x == 1.0. If this is not possible, then the tests should be done replacing tests like x == 1.0 with abs(x - 1.0) < eps. Function abs() returns the absolute value, in simpler words, makes all values positive or zero, by changing the sign of negative values, or in mathematical notation $|x| = | - x|$.

```
a == 0.0 # may not always work
##  [1] FALSE FALSE FALSE FALSE FALSE FALSE FALSE FALSE FALSE FALSE

abs(a) < 1e-15 # is safer
##  [1] FALSE FALSE FALSE FALSE FALSE FALSE FALSE FALSE FALSE FALSE


sin(pi) == 0.0 # angle in radians, not degrees!
## [1] FALSE

sin(2 * pi) == 0.0
## [1] FALSE

abs(sin(pi)) < 1e-15
## [1] TRUE

abs(sin(2 * pi)) < 1e-15
## [1] TRUE

sin(pi)
## [1] 1.224606e-16

sin(2 * pi)
## [1] -2.449213e-16
```

2.6 Sets and set operations

The R language supports set operations on vectors. They can be useful in many different contexts when manipulating and comparing vectors of values. In Bioinformatics it is usual, for example, to have character vectors of gene tags. We may have a vector for each of a set of different samples, and need to compare them. However, we start by using a more mundane example, everyday shopping.

```
fruits <- c("apple", "pear", "orange", "lemon", "tangerine")
bakery <- c("bread", "buns", "cake", "cookies")
dairy <- c("milk", "butter", "cheese")
shopping <- c("bread", "butter", "apple", "cheese", "orange")
intersect(fruits, shopping)
## [1] "apple"  "orange"

intersect(bakery, shopping)
## [1] "bread"
```

```
intersect(dairy, shopping)
## [1] "butter" "cheese"

"lemon" %in% dairy
## [1] FALSE

"lemon" %in% fruits
## [1] TRUE

setdiff(union(bakery, dairy), shopping)
## [1] "buns"    "cake"    "cookies" "milk"
```

We continue next with abstract (symbolic) examples.

```
my.set <- c("a", "b", "c", "b")
```

To test if a given value belongs to a set, we use operator `%in%`. In the algebra of sets notation, this is written $a \in A$, where A is a set and a a member. The second statement shows that the `%in%` operator is vectorized on its left-hand-side (lhs) operand, returning a logical vector.

```
"a" %in% my.set
## [1] TRUE

c("a", "a", "z") %in% my.set
## [1]  TRUE  TRUE FALSE
```

The negation of inclusion is $a \notin A$, and coded in R by applying the negation operator `!` to the result of the test done with `%in%`.

```
!"a" %in% my.set
## [1] FALSE

!c("a", "a", "z") %in% my.set
## [1] FALSE FALSE  TRUE
```

Although inclusion is a set operation, it is also very useful for the simplification of `if()`…`else` statements by replacing multiple tests for alternative constant values of the same `mode` chained by multiple `|` operators.

> Use operator `%in%` to simplify the following comparison.
>
> ```
> x <- c("a", "a", "z")
> x == "a" | x == "b" | x == "c" | x == "d"
> ```

With `unique()` we convert a vector of possibly repeated values into a set of unique values. In the algebra of sets, a certain object belongs or not to a set. Consequently, in a set, multiple copies of the same object or value are meaningless.

```
unique(my.set)
## [1] "a" "b" "c"

c("a", "a", "z") %in% unique(my.set)
## [1]  TRUE  TRUE FALSE
```

In the notation used in algebra of sets, the set union operator is $\cup$ while the intersection operator is $\cap$. If we have sets A and B, their union is given by $A \cup B$—in the next three examples, `c("a", "a", "z")` is a constant, while `my.set` is a variable.

```
union(c("a", "a", "z"), my.set)
## [1] "a" "z" "b" "c"
```

If we have sets A and B, their intersection is given by $A \cap B$.

```
intersect(c("a", "a", "z"), my.set)
## [1] "a"
```

What do you expect to be the difference between the values returned by the three statements in the code chunk below? Before running them, write down your expectations about the value each one will return. Only then run the code. Independently of whether your predictions were correct or not, write down an explanation of what each statement's operation is.

```
union(c("a", "a", "z"), my.set)
c(c("a", "a", "z"), my.set)
c("a", "a", "z", my.set)
```

In the algebra of sets notation $A \subseteq B$, where A and B are sets, indicates that A is a subset or equal to B. For a true subset, the notation is $A \subset B$. The operators with the reverse direction are $\supseteq$ and $\supset$. Implement these four operations in four R statements, and test them on sets (represented by R vectors) with different "overlap" among set members.

All set algebra examples above use character vectors and character constants. This is just the most frequent use case. Sets operations are valid on vectors of any atomic class, including `integer`, and computed values can be part of statements. In the second and third statements in the next chunk, we need to use additional parentheses to alter the default order of precedence between arithmetic and set operators.

```
9L %in% 2L:4L
## [1] FALSE

9L %in% ((2L:4L) * (2L:4L))
## [1] TRUE

c(1L, 16L) %in% ((2L:4L) * (2L:4L))
## [1] FALSE  TRUE
```

Empty sets are an important component of the algebra of sets, in R they are represented as vectors of zero length. Vectors and lists of zero length, which the R language fully supports, can be used to "encode" emptiness also in other contexts. These vectors do belong to a class such as numeric or character and must be compatible with other operands in an expression. By default, constructors for vectors, construct empty vectors.

```
length(integer())
## [1] 0

1L %in% integer()
## [1] FALSE

setdiff(1L:4L, union(1L:4L, integer()))
## integer(0)
```

Although set operators are defined for numeric vectors, rounding errors in 'floats' can result in unexpected results (see section 2.5 on page 33). The next two examples do, however, return the correct answers.

```
9 %in% (2:4)^2
## [1] TRUE

c(1, 5) %in% (1:10)^2
## [1]  TRUE FALSE
```

2.7   Character values

Character variables can be used to store any character. Character constants are written by enclosing characters in quotes. There are three types of quotes in the ASCII character set, double quotes ", single quotes ', and back ticks `. The first two types of quotes can be used as delimiters of character constants.

```
a <- "A"
a
## [1] "A"
```

```
b <- 'A'
b
## [1] "A"

a == b
## [1] TRUE
```

> ☕ In many computer languages, vectors of characters are distinct from vectors of character strings. In these languages, character vectors store at each index position a single character, while vectors of character strings store at each index position strings of characters of various lengths, such as words or sentences. If you are familiar with C or C++, you need to keep in mind that C's char and R's character are not equivalent and that in R, character vectors are vectors of character strings. In contrast to these other languages, in R there is no predefined class for vectors of individual characters and character constants enclosed in double or single quotes are not different.

Concatenating character vectors of length one does not yield a longer character string, it yields instead a longer vector.

```
a <- 'A'
b <- "bcdefg"
c <- "123"
d <- c(a, b, c)
d
## [1] "A"       "bcdefg" "123"
```

Having two different delimiters available makes it possible to choose the type of quotes used as delimiters so that other quotes can be included in a string.

```
a <- "He said 'hello' when he came in"
a
## [1] "He said 'hello' when he came in"

b <- 'He said "hello" when he came in'
b
## [1] "He said \"hello\" when he came in"
```

The outer quotes are not part of the string, they are "delimiters" used to mark the boundaries. As you can see when b is printed special characters can be represented using "escape sequences". There are several of them, and here we will show just four, new line (\n) and tab (\t), \" the escape code for a quotation mark within a string and \\ the escape code for a single backslash \. We also show here the different behavior of print() and cat(), with cat() *interpreting* the escape sequences and print() displaying them as entered.

```
c <- "abc\ndef\tx\"yz\"\\\tm"
print(c)
## [1] "abc\ndef\tx\"yz\"\\\tm"

cat(c)
## abc
## def x"yz"\ m
```

The *escape codes* work only in some contexts, as when using cat() to generate the output. For example, the new-line escape (\n) can be embedded in strings used for axis-label, title or label in a plot to split them over two or more lines.

2.8   The 'mode' and 'class' of objects

Variables have a *mode* that depends on what is stored in them. But different from other languages, assignment to a variable of a different mode is allowed and in most cases its mode changes together with its contents. However, there is a restriction that all elements in a vector, array or matrix, must be of the same mode. While this is not required for lists, which can be heterogenous. In practice this means that we can assign an object, such as a vector, with a different mode to a name already in use, but we cannot use indexing to assign an object of a different mode to individual members of a vector, matrix or array. Functions with names starting with is. are tests returning a logical value, TRUE, FALSE or NA. Function mode() returns the mode of an object, as a character string and typeof() returns R's internal type or storage mode.

```
my_var <- 1:5
mode(my_var) # no distinction of integer or double
## [1] "numeric"

typeof(my_var)
## [1] "integer"

is.numeric(my_var) # no distinction of integer or double
## [1] TRUE

is.double(my_var)
## [1] FALSE

is.integer(my_var)
## [1] TRUE

is.logical(my_var)
## [1] FALSE

is.character(my_var)
## [1] FALSE

my_var <- "abc"
mode(my_var)
## [1] "character"
```

While *mode* is a fundamental property, and limited to those modes defined as part of the R language, the concept of *class*, is different in that new classes can be defined in user code. In particular, different R objects of a given mode, such as numeric, can belong to different classes. The use of classes for dispatching functions is discussed in section 5.4 on page 172, in relation to object-oriented programming in R. Method class() is used to query the class of an object, and method `inherits()` is used to test if an object belongs to a specific class or not (including "parent" classes, to be later described).

```
class(my_var)
## [1] "character"

inherits(my_var, "character")
## [1] TRUE

inherits(my_var, "numeric")
## [1] FALSE
```

2.9 'Type' conversions

The least-intuitive type conversions are those related to logical values. All others are as one would expect. By convention, functions used to convert objects from one mode to a different one have names starting with `as.`[1].

```
as.character(1)
## [1] "1"

as.numeric("1")
## [1] 1

as.logical("TRUE")
## [1] TRUE

as.logical("NA")
## [1] NA
```

Conversion takes place automatically in arithmetic and logical expressions.

```
TRUE + 10
## [1] 11

1 || 0
## [1] TRUE

FALSE | -2:2
## [1]  TRUE  TRUE FALSE  TRUE  TRUE
```

> There is some flexibility in the conversion from character strings into `numeric` and `logical` values. Use the examples below plus your own variations to get an idea of what strings are acceptable and correctly converted and which are not. Do also pay attention at the conversion between `numeric` and `logical` values.

---

[1] Except for some packages in the 'tidyverse' that use names starting with `as_` instead of `as.`.

```
as.character(3.0e10)
as.numeric("5E+5")
as.numeric("A")
as.numeric(TRUE)
as.numeric(FALSE)
as.logical("T")
as.logical("t")
as.logical("true")
as.logical(100)
as.logical(0)
as.logical(-1)
```

Compare the values returned by trunc() and as.integer() when applied to a floating point number, such as 12.34. Check for the equality of values, and for the *class* of the returned objects.

Using conversions, the difference between the length of a character vector and the number of characters composing each member "string" within a vector is obvious.

```
f <- c("1", "2", "3")
length(f)
## [1] 3

g <- "123"
length(g)
## [1] 1

as.numeric(f)
## [1] 1 2 3

as.numeric(g)
## [1] 123
```

Other functions relevant to the "conversion" of numbers and other values are format(), and sprintf(). These two functions return character strings, instead of numeric or other values, and are useful for printing output. One could think of these functions as advanced conversion functions returning formatted, and possibly combined and annotated, character strings. However, they are usually not considered normal conversion functions, as they are very rarely used in a way that preserves the original precision of the input values. We show here the use of format() and sprintf() with numeric values, but they can also be used with values of other modes.

When using format(), the format used to display numbers is set by passing arguments to several different parameters. As print() calls format() to make numbers *pretty* it accepts the same options.

```
x = c(123.4567890, 1.0)
format(x) # using defaults
## [1] "123.4568" "  1.0000"

format(x[1]) # using defaults
## [1] "123.4568"

format(x[2]) # using defaults
## [1] "1"

format(x, digits = 3, nsmall = 1)
## [1] "123.5" "  1.0"

format(x[1], digits = 3, nsmall = 1)
## [1] "123.5"

format(x[2], digits = 3, nsmall = 1)
## [1] "1.0"

format(x, digits = 3, scientific = TRUE)
## [1] "1.23e+02" "1.00e+00"
```

Function `sprintf()` is similar to C's function of the same name. The user interface is rather unusual, but very powerful, once one learns the syntax. All the formatting is specified using a `character` string as template. In this template, placeholders for data and the formatting instructions are embedded using special codes. These codes start with a percent character. We show in the example below the use of some of these: `f` is used for `numeric` values to be formatted according to a "fixed point," while `g` is used when we set the number of significant digits and `e` for exponential or *scientific* notation.

```
x = c(123.4567890, 1.0)
sprintf("The numbers are: %4.2f and %.0f", x[1], x[2])
## [1] "The numbers are: 123.46 and 1"

sprintf("The numbers are: %.4g and %.2g", x[1], x[2])
## [1] "The numbers are: 123.5 and 1"

sprintf("The numbers are: %4.2e and %.0e", x[1], x[2])
## [1] "The numbers are: 1.23e+02 and 1e+00"
```

In the template `"The numbers are: %4.2f and %.0f"`, there are two placeholders for `numeric` values, `%4.2f` and `%.0f`, so in addition to the template, we pass two values extracted from the first two positions of vector `x`. These could have been two different vectors of length one, or even numeric constants. The template itself does not need to be a `character` constant as in these examples, as a variable can be also passed as argument.

Function `format()` may be easier to use, in some cases, but `sprintf()` is more flexible and powerful. Those with experience in the use of the C language will already know about `sprintf()` and its use of templates for formatting output. Even if you are familiar with C, look up the help pages for both functions, and practice, by trying to create the same formatted output by means of the two functions. Do also play with these functions with other types of data like `integer` and `character`.

☕ We have above described NA as a single value ignoring modes, but in reality NA s come in various flavors. NA_real_, NA_character_, etc. and NA defaults to an NA of class `logical`. NA is normally converted on the fly to other modes when needed, so in general NA is all we need to use.

```
a <- c(1, NA)
is.numeric(a[2])
## [1] TRUE

is.numeric(NA)
## [1] FALSE

b <- c("abc", NA)
is.character(b[2])
## [1] TRUE

is.character(NA)
## [1] FALSE

class(NA)
## [1] "logical"
```

Even the statement below works transparently.

```
a[3] <- b[2]
```

2.10 Vector manipulation

If you have read earlier sections of this chapter, you already know how to create a vector. R's vectors are equivalent to what would be written in mathematical notation as $x_{1...n} = a_1, a_2, \ldots, a_i, \ldots, a_n$, they are not the equivalent to the vectors, common in Physics, which are symbolized with an arrow as an "accent," such as $\vec{F}$.

In this section we are going to see how to extract or retrieve, replace, and move elements such as a_2 from a vector. Elements are extracted using an index enclosed in single square brackets. The index indicates the position in the vector, starting from one, following the usual mathematical tradition. What in maths would be a_i for a vector $a_{1...n}$, in R is represented as a[i] and the whole vector as earlier seen as a.

```
a <- letters[1:10]
a
##  [1] "a" "b" "c" "d" "e" "f" "g" "h" "i" "j"
```

```
a[2]
## [1] "b"
```

⌨ Four constant vectors are available in R: letters, LETTERS, month.name and month.abb, of which we used letters in the example above. These vectors are always for English, irrespective of the locale.

⚠ In R, indexes always start from one, while in some other programming languages such as C and C++, indexes start from zero. It is important to be aware of this difference, as many computation algorithms are valid only under a given indexing convention.

It is possible to extract a subset of the elements of a vector in a single operation, using a vector of indexes. The positions of the extracted elements in the result ("returned value") are determined by the ordering of the members of the vector of indexes—easier to demonstrate than to explain.

```
a[c(3,2)]
## [1] "c" "b"
```

```
a[10:1]
##  [1] "j" "i" "h" "g" "f" "e" "d" "c" "b" "a"
```

▲ The length of the indexing vector is not restricted by the length of the indexed vector. However, only numerical indexes that match positions present in the indexed vector can extract values. Those values in the indexing vector pointing to positions that are not present in the indexed vector, result in NAS. This is easier to learn by *playing* with R, than from explanations. Play with R, using the following examples as a starting point.

```
length(a)
a[c(3,3,3,3)]
a[c(10:1, 1:10)]
a[c(1,11)]
a[11]
```

Have you tried some of your own examples? If not yet, do *play* with additional variations of your own before continuing.

Negative indexes have a special meaning; they indicate the positions at which values should be excluded. Be aware that it is *illegal* to mix positive and negative values in the same indexing operation.

```
a[-2]
## [1] "a" "c" "d" "e" "f" "g" "h" "i" "j"

a[-c(3,2)]
## [1] "a" "d" "e" "f" "g" "h" "i" "j"

a[-3:-2]
## [1] "a" "d" "e" "f" "g" "h" "i" "j"
```

Results from indexing with special values and zero may be surprising. Try to build a rule from the examples below, a rule that will help you remember what to expect next time you are confronted with similar statements using "subscripts" which are special values instead of integers larger or equal to one—this is likely to happen sooner or later as these special values can be returned by different R expressions depending on the value of operands or function arguments, some of them described earlier in this chapter.

```
a[ ]
a[0]
a[numeric(0)]
a[NA]
a[c(1, NA)]
a[NULL]
a[c(1, NULL)]
```

Another way of indexing, which is very handy, but not available in most other programming languages, is indexing with a vector of logical values. The logical vector used for indexing is usually of the same length as the vector from which elements are going to be selected. However, this is not a requirement, because if the logical vector of indexes is shorter than the indexed vector, it is "recycled" as discussed above in relation to other operators.

```
a[TRUE]
##  [1] "a" "b" "c" "d" "e" "f" "g" "h" "i" "j"

a[FALSE]
## character(0)

a[c(TRUE, FALSE)]
## [1] "a" "c" "e" "g" "i"

a[c(FALSE, TRUE)]
## [1] "b" "d" "f" "h" "j"

a > "c"
##  [1] FALSE FALSE FALSE  TRUE  TRUE  TRUE  TRUE  TRUE  TRUE  TRUE

a[a > "c"]
## [1] "d" "e" "f" "g" "h" "i" "j"
```

Indexing with logical vectors is very frequently used in R because comparison operators are vectorized. Comparison operators, when applied to a vector, return a `logical` vector, a vector that can be used to extract the elements for which the result of the comparison test was TRUE.

The examples in this text box demonstrate additional uses of logical vectors: 1) the logical vector returned by a vectorized comparison can be stored in a variable, and the variable used as a "selector" for extracting a subset of values from the same vector, or from a different vector.

```
a <- letters[1:10]
b <- 1:10
selector <- a > "c"
selector
a[selector]
b[selector]
```

Numerical indexes can be obtained from a logical vector by means of function `which()`.

```
indexes <- which(a > "c")
indexes
a[indexes]
b[indexes]
```

Make sure to understand the examples above. These constructs are very widely used in R because they allow for concise code that is easy to understand once you are familiar with the indexing rules. However, if you do not command these rules, many of these terse statements will be unintelligible to you.

Indexing can be used on either side of an assignment expression. In the chunk below, we use the extraction operator on the left-hand side of the assignments to replace values only at selected positions in the vector. This may look rather esoteric at first sight, but it is just a simple extension of the logic of indexing described above. It works, because the low precedence of the <- operator results in both the left-hand side and the right-hand side being fully evaluated before the assignment takes place. To make the changes to the vectors easier to follow, we use identical vectors with different names for each of these examples.

```
a <- 1:10
a
## [1]  1  2  3  4  5  6  7  8  9 10

a[1] <- 99
a
## [1] 99  2  3  4  5  6  7  8  9 10

b <- 1:10
b[c(2,4)] <- -99 # recycling
b
## [1]   1 -99   3 -99   5   6   7   8   9  10
```

```
c <- 1:10
c[c(2,4)] <- c(-99, 99)
c
## [1]   1 -99   3  99   5   6   7   8   9  10

d <- 1:10
d[TRUE] <- 1 # recycling
d
## [1] 1 1 1 1 1 1 1 1 1 1

e <- 1:10
e <- 1  # no recycling
e
## [1] 1
```

We can also use subscripting on both sides of the assignment operator, for example, to swap two elements.

```
a <- letters[1:10]
a[1:2] <- a[2:1]
a
## [1] "b" "a" "c" "d" "e" "f" "g" "h" "i" "j"
```

Do play with subscripts to your heart's content, really grasping how they work and how they can be used, will be very useful in anything you do in the future with R. Even the contrived example below follows the same simple rules, just study it bit by bit. Hint: the second statement in the chunk below, modifies a, so, when studying variations of this example you will need to recreate a by executing the first statement, each time you run a variation of the second statement.

```
a <- letters[1:10]
a[5:1] <- a[c(TRUE,FALSE)]
a
## [1] "i" "g" "e" "c" "a" "f" "g" "h" "i" "j"
```

In R, indexing with positional indexes can be done with `integer` or `numeric` values. Numeric values can be floats, but for indexing, only integer values are meaningful. Consequently, `double` values are converted into `integer` values when used as indexes. The conversion is done invisibly, but it does slow down computations slightly. When working on big data sets, explicitly using `integer` values can improve performance.

```
b <- LETTERS[1:10]
b[1]
## [1] "A"

b[1.1]
## [1] "A"

b[1.9999] # surprise!!
## [1] "A"

b[2]
## [1] "B"
```

From this experiment, we can learn that if positive indexes are not whole numbers, they are truncated to the next smaller integer.

```
b <- LETTERS[1:10]
b[-1]
## [1] "B" "C" "D" "E" "F" "G" "H" "I" "J"

b[-1.1]
## [1] "B" "C" "D" "E" "F" "G" "H" "I" "J"

b[-1.9999]
## [1] "B" "C" "D" "E" "F" "G" "H" "I" "J"

b[-2]
## [1] "A" "C" "D" "E" "F" "G" "H" "I" "J"
```

From this experiment, we can learn that if negative indexes are not whole numbers, they are truncated to the next larger (less negative) integer. In conclusion, `double` index values behave as if they where sanitized using function `trunc()`.

This example also shows how one can tease out of R its rules through experimentation.

A frequent operation on vectors is sorting them into an increasing or decreasing order. The most direct approach is to use `sort()`.

```
my.vector <- c(10, 4, 22, 1, 4)
sort(my.vector)
## [1]  1  4  4 10 22

sort(my.vector, decreasing = TRUE)
## [1] 22 10  4  4  1
```

An indirect way of sorting a vector, possibly based on a different vector, is to generate with `order()` a vector of numerical indexes that can be used to achieve the ordering.

```
order(my.vector)
## [1] 4 2 5 1 3
```

```
my.vector[order(my.vector)]
## [1]  1  4  4 10 22

another.vector <- c("ab", "aa", "c", "zy", "e")
another.vector[order(my.vector)]
## [1] "zy" "aa" "e"  "ab" "c"
```

☕ A problem linked to sorting that we may face is counting how many copies of each value are present in a vector. We need to use two functions `sort()` and `rle()`. The second of these functions computes *run length* as used in *run length encoding* for which *rle* is an abbreviation. A *run* is a series of consecutive identical values. As the objective is to count the number of copies of each value present, we need first to sort the vector.

```
my.letters <- letters[c(1,5,10,3,1,4,21,1,10)]
my.letters
## [1] "a" "e" "j" "c" "a" "d" "u" "a" "j"

sort(my.letters)
## [1] "a" "a" "a" "c" "d" "e" "j" "j" "u"

rle(sort(my.letters))
## Run Length Encoding
##   lengths: int [1:6] 3 1 1 1 2 1
##   values : chr [1:6] "a" "c" "d" "e" "j" "u"
```

The second and third statements are only to demonstrate the effect of each step. The last statement uses nested function calls to compute the number of copies of each value in the vector.

2.11 Matrices and multidimensional arrays

Vectors have a single dimension, and, as we saw above, we can query their length with method `length()`. Matrices have two dimensions, which can be queried with `dim()`, `ncol()` and `nrow()`. R arrays can have any number of dimensions, even a single dimension, which can be queried with method `dim()`. As expected `is.vector()`, `is.matrix()` and `is.array()` can be used to query the class.

We can create a new matrix using the `matrix()` or `as.matrix()` constructors. The first argument of `matrix()` is a vector. In the same way as vectors, matrices are homogeneous, all elements are of the same type.

```
matrix(1:15, ncol = 3)
##      [,1] [,2] [,3]
## [1,]    1    6   11
## [2,]    2    7   12
```

```
## [3,]    3    8    13
## [4,]    4    9    14
## [5,]    5   10    15
```

```
matrix(1:15, nrow = 3)
##      [,1] [,2] [,3] [,4] [,5]
## [1,]    1    4    7   10   13
## [2,]    2    5    8   11   14
## [3,]    3    6    9   12   15
```

When a vector is converted to a matrix, R's default is to allocate the values in the vector to the matrix starting from the leftmost column, and within the column, down from the top. Once the first column is filled, the process continues from the top of the next column, as can be seen above. This order can be changed as you will discover in the playground below.

Check in the help page for the `matrix` constructor how to use the `byrow` parameter to alter the default order in which the elements of the vector are allocated to columns and rows of the new matrix.

```
help(matrix)
```

While you are looking at the help page, also consider the default number of columns and rows.

```
matrix(1:15)
```

And to start getting a sense of how to interpret error and warning messages, run the code below and make sure you understand which problem is being reported. Before executing the statement, analyze it and predict what the returned value will be. Afterwards, compare your prediction, to the value actually returned.

```
matrix(1:15, ncol = 2)
```

Subscripting of matrices and arrays is consistent with that used for vectors; we only need to supply an indexing vector, or leave a blank space, for each dimension. A matrix has two dimensions, so to access any element or group of elements, we use two indices. The only complication is that there are two possible orders in which, in principle, indexes could be supplied. In R, indexes for matrices are written "row first." In simpler words, the first index value selects rows, and the second one, columns.

```
A <- matrix(1:20, ncol = 4)
A
##      [,1] [,2] [,3] [,4]
## [1,]    1    6   11   16
## [2,]    2    7   12   17
## [3,]    3    8   13   18
```

```
## [4,]    4    9   14   19
## [5,]    5   10   15   20

A[1, 1]
## [1] 1
```

Remind yourself of how indexing of vectors works in R (see section 2.10 on page 45). We will now apply the same rules in two dimensions.

```
A[1, ]
## [1]  1  6 11 16

A[ , 1]
## [1] 1 2 3 4 5

A[2:3, c(1,3)]
##      [,1] [,2]
## [1,]    2   12
## [2,]    3   13

A[3, 4] <- 99
A
##      [,1] [,2] [,3] [,4]
## [1,]    1    6   11   16
## [2,]    2    7   12   17
## [3,]    3    8   13   99
## [4,]    4    9   14   19
## [5,]    5   10   15   20

A[4:3, 2:1] <- A[3:4, 1:2]
A
##      [,1] [,2] [,3] [,4]
## [1,]    1    6   11   16
## [2,]    2    7   12   17
## [3,]    9    4   13   99
## [4,]    8    3   14   19
## [5,]    5   10   15   20
```

☕ In R, a matrix can have a single row, a single column, a single element or no elements. However, in all cases, a matrix will have *dimensions* of length two defined and stored as an attribute.

```
my.vector <- 1:6
dim(my.vector)
## NULL
```

```
one.col.matrix <- matrix(1:6, ncol = 1)
dim(one.col.matrix)
## [1] 6 1

two.col.matrix <- matrix(1:6, ncol = 2)
dim(two.col.matrix)
## [1] 3 2

one.elem.matrix <- matrix(1, ncol = 1)
dim(one.elem.matrix)
## [1] 1 1

no.elem.matrix <- matrix(numeric(), ncol = 0)
dim(no.elem.matrix)
## [1] 0 0
```

Arrays are similar to matrices, but can have more than two dimensions, which are specified with the `dim` argument to the `array()` constructor.

```
B <- array(1:27, dim = c(3, 3, 3))
B
## , , 1
##
##      [,1] [,2] [,3]
## [1,]    1    4    7
## [2,]    2    5    8
## [3,]    3    6    9
##
## , , 2
##
##      [,1] [,2] [,3]
## [1,]   10   13   16
## [2,]   11   14   17
## [3,]   12   15   18
##
## , , 3
##
##      [,1] [,2] [,3]
## [1,]   19   22   25
## [2,]   20   23   26
## [3,]   21   24   27

B[2, 2, 2]
## [1] 14
```

In the chunk above, the length of the supplied vector is the product of the dimensions, $27 = 3 \times 3 \times 3$.

How do you use indexes to extract the second element of the original vector, in each of the following matrices and arrays?

```
v <- 1:10
m2c <- matrix(v, ncol = 2)
m2cr <- matrix(v, ncol = 2, byrow = TRUE)
m2r <- matrix(v, nrow = 2)
m2rc <- matrix(v, nrow = 2, byrow = TRUE)


v <- 1:10
a2c <- array(v, dim = c(5, 2))
a2c <- array(v, dim = c(5, 2), dimnames = list(NULL, c("c1", "c2")))
a2r <- array(v, dim = c(2, 5))
```

Be aware that vectors and one-dimensional arrays are not the same thing, while two-dimensional arrays are matrices.

1. Use the different constructors and query methods to explore this, and its consequences.

2. Convert a matrix into a vector using `unlist()` and `as.vector()` and compare the returned values.

Operators for matrices are available in R, as matrices are used in many statistical algorithms. We will not describe them all here, only `t()` and some specializations of arithmetic operators. Function `t()` transposes a matrix, by swapping columns and rows.

```
A <- matrix(1:20, ncol = 4)
A
##      [,1] [,2] [,3] [,4]
## [1,]    1    6   11   16
## [2,]    2    7   12   17
## [3,]    3    8   13   18
## [4,]    4    9   14   19
## [5,]    5   10   15   20

t(A)
##      [,1] [,2] [,3] [,4] [,5]
## [1,]    1    2    3    4    5
## [2,]    6    7    8    9   10
## [3,]   11   12   13   14   15
## [4,]   16   17   18   19   20
```

As with vectors, recycling applies to arithmetic operators when applied to matrices.

```
A + 2
##      [,1] [,2] [,3] [,4]
## [1,]    3    8   13   18
## [2,]    4    9   14   19
## [3,]    5   10   15   20
## [4,]    6   11   16   21
## [5,]    7   12   17   22

A * 0:1
```

```
##      [,1] [,2] [,3] [,4]
## [1,]    0    6    0   16
## [2,]    2    0   12    0
## [3,]    0    8    0   18
## [4,]    4    0   14    0
## [5,]    0   10    0   20
```

```
A * 1:0
##      [,1] [,2] [,3] [,4]
## [1,]    1    0   11    0
## [2,]    0    7    0   17
## [3,]    3    0   13    0
## [4,]    0    9    0   19
## [5,]    5    0   15    0
```

In the examples above with the usual multiplication operator *, the operation described is not a matrix product, but instead, the products between individual elements of the matrix and vectors. Matrix multiplication is indicated by operator %*%.

```
B <- matrix(1:16, ncol = 4)
B * B
##      [,1] [,2] [,3] [,4]
## [1,]    1   25   81  169
## [2,]    4   36  100  196
## [3,]    9   49  121  225
## [4,]   16   64  144  256
```

```
B %*% B
##      [,1] [,2] [,3] [,4]
## [1,]   90  202  314  426
## [2,]  100  228  356  484
## [3,]  110  254  398  542
## [4,]  120  280  440  600
```

Other operators and functions for matrix algebra like cross-product (crossprod()), extracting or replacing the diagonal (diag()) are available in base R. Packages, including 'matrixStats', provide additional functions and operators for matrices.

2.12   Factors

Factors are used to indicate categories, most frequently the factors describing the treatments in an experiment, or categories in a survey. They can be created either from numerical or character vectors. The different possible values are called *levels*. Normal factors created with factor() are unordered or categorical. R also supports ordered factors that can be created with function ordered().

```
my.vector <- c("treated", "treated", "control", "control", "control", "treated")
my.factor <- factor(my.vector)
my.factor
```

```
## [1] treated treated control control control treated
## Levels: control treated

my.factor <- factor(x = my.vector, levels = c("treated", "control"))
my.factor
## [1] treated treated control control control treated
## Levels: treated control
```

The labels ("names") of the levels can be set when the factor is created. In this case, when calling `factor()`, parameters `levels` and `labels` should both be passed a vector as argument, with levels and matching labels in the same position in the two vectors. The argument passed to `levels` determines the order of the levels based on their old names or values, and the argument passed to `labels` gives new names to the levels.

```
my.vector <- c(1, 1, 0, 0, 0, 1)
my.factor <- factor(x = my.vector, levels = c(1, 0), labels = c("treated", "control"))
my.factor
## [1] treated treated control control control treated
## Levels: treated control
```

It is always preferable to use meaningful labels for levels, although it is also possible to use numbers.

In the examples above we passed a numeric vector or a character vector as an argument for parameter x of function `factor()`. It is also possible to pass a `factor` as an argument for parameter x. We use indexing with a test returning a logical vector to extract all "controls." We use function `levels()` to look at the levels of the factors.

```
levels(my.factor)
## [1] "treated" "control"

control.factor <- my.factor[my.factor == "control"]
control.factor
## [1] control control control
## Levels: treated control

control.factor <- factor(control.factor)
control.factor
## [1] control control control
## Levels: control
```

It can be seen above that subsetting does not drop unused factor levels, and that `factor()` can be used to explicitly drop the unused factor levels.

When the pattern of levels is regular, it is possible to use function `gl()` to *generate levels* in a factor. Nowadays, it is more usual to read data into R from files in which the treatment codes are already available as character strings or numeric values, however, when we need to create a factor within R, `gl()` can save some typing.

```
gl(2, 5, labels = c("A", "B"))
## [1] A A A A A B B B B B
## Levels: A B
```

Converting factors into numbers is not intuitive, even in the case where a factor was created from a `numeric` vector.

```
my.vector2 <- rep(3:5, 4)
my.vector2
## [1] 3 4 5 3 4 5 3 4 5 3 4 5
```

```
my.factor2 <- factor(my.vector2)
my.factor2
## [1] 3 4 5 3 4 5 3 4 5 3 4 5
## Levels: 3 4 5
```

```
as.numeric(my.factor2)
## [1] 1 2 3 1 2 3 1 2 3 1 2 3
```

```
as.numeric(as.character(my.factor2))
## [1] 3 4 5 3 4 5 3 4 5 3 4 5
```

Why is a double conversion needed? Internally, factor levels are stored as running integers starting from one, and those are the numbers returned by `as.numeric()` when applied to a factor. The labels of the factor levels are always stored as character strings, even when these characters are digits. In contrast to `as.numeric()`, `as.character()` returns the character labels of the levels. If these character strings represent numbers, they can be converted, in a second step, using `as.numeric()` into the original numeric values.

```
class(my.factor2)
## [1] "factor"
```

```
mode(my.factor2)
## [1] "numeric"
```

Create a factor with levels labeled with words. Create another factor with the levels labeled with the same words, but ordered differently. After this convert both factors to numeric vectors using `as.numeric()`. Explain why the two numeric vectors differ or not from each other.

Factors are very important in R. In contrast to other statistical software in which the role of a variable is set when defining a model to be fitted or when setting up a test, in R, models are specified exactly in the same way for ANOVA and regression analysis, both as *linear models*. The type of model that is fitted is decided by whether the explanatory variable is a factor (giving ANOVA) or a numerical variable (giving regression). This makes a lot of sense, because in most cases, considering an explanatory variable as categorical or not, depends on the design of the experiment or survey, and in other words, is a property of the data and the experiment or survey that gave origin to them, rather than of the data analysis.

The order of the levels in a `factor` does not affect simple calculations or the values plotted, but it does affect how the output is printed, the order of the levels in the scales of plots, and in some cases the contrasts in significance tests. The default ordering is alphabetical, and is established at the time a factor is created. Consequently, rather frequently the default ordering of levels is not the one needed. As shown above, parameter `levels` in the constructor makes it possible to set the order of the levels. It is also possible to change the ordering of an existing factor.

> **Renaming factor levels.** The most direct way is using `levels()<-` as shown below, but it is also possible to use `factor()`. The difference is that `factor()` drops the unused levels and `levels()` only renames existing levels, all of them by position. (Although we here use `character` strings that are only one character long, longer strings can be set as labels in exactly the same way.
>
> ```
> my.factor1 <- gl(4, 3, labels = c("A", "F", "B", "Z"))
> my.factor1
> ## [1] A A A F F F B B B Z Z Z
> ## Levels: A F B Z
>
> levels(my.factor1) <- c("a", "b", "c", "d")
> my.factor1
> ## [1] a a a b b b c c c d d d
> ## Levels: a b c d
> ```
>
> Or more safely by matching names—i.e., order in the list of replacement values is irrelevant.
>
> ```
> my.factor1 <- gl(4, 3, labels = c("A", "F", "B", "Z"))
> my.factor1
> ## [1] A A A F F F B B B Z Z Z
> ## Levels: A F B Z
>
> levels(my.factor1) <- list("a" = "A", "d" = "Z", "c" = "B", "b" = "F")
> my.factor1
> ## [1] a a a b b b c c c d d d
> ## Levels: a d c b
> ```
>
> Or alternatively by position and replacing the labels of only some levels—i.e., rather unsafe.
>
> ```
> my.factor1 <- gl(4, 3, labels = c("A", "F", "B", "Z"))
> my.factor1
> ## [1] A A A F F F B B B Z Z Z
> ## Levels: A F B Z
> ```

```
levels(my.factor1)[c(1, 4)] <- c("a", "d")
my.factor1
## [1] a a a F F F B B B d d d
## Levels: a F B d
```

☕ **Merging factor levels.** We use `factor()` as shown below, setting the same label for the levels we want to merge.

```
my.factor1 <- gl(4, 3, labels = c("A", "F", "B", "Z"))
my.factor1
## [1] A A A F F F B B B Z Z Z
## Levels: A F B Z

factor(my.factor1,
       levels = c("A", "B", "F", "Z"),
       labels = c("A", "B", "C", "C"))
## [1] A A A C C C B B B C C C
## Levels: A B C
```

☕ **Reordering factor levels.** The simplest approach is to use `factor()` and its `levels` parameter. The only complication is that the names of the existing levels and those passed as an argument need to match, and typing mistakes can cause bugs. To avoid the error-prone step, in all examples except the first, we use `levels()` to retrieve the names of the levels from the factor itself.

```
levels(my.factor2)
## [1] "3" "4" "5"

my.factor2 <- factor(my.factor2, levels = c("5", "3", "4"))
levels(my.factor2)
## [1] "5" "3" "4"

my.factor2 <- factor(my.factor2, levels = rev(levels(my.factor2)))
levels(my.factor2)
## [1] "4" "3" "5"


my.factor2 <- factor(my.factor2,
                     levels = sort(levels(my.factor2), decreasing = TRUE))
levels(my.factor2)
## [1] "5" "4" "3"

my.factor2 <- factor(my.factor2, levels = levels(my.factor2)[c(2, 1, 3)])
levels(my.factor2)
## [1] "4" "5" "3"
```

Reordering the levels of a factor based on summary quantities from data is very useful, especially when plotting. Function `reorder()` can be used in this case. It defaults to using `mean()` for summaries, but other suitable functions can be supplied in its place.

```
my.factor3 <- gl(2, 5, labels = c("A", "B"))
my.vector3 <- c(5.6, 7.3, 3.1, 8.7, 6.9, 2.4, 4.5, 2.1, 1.4, 2.0)
levels(my.factor3)
## [1] "A" "B"

my.factor3ord <- reorder(my.factor3, my.vector3)
levels(my.factor3ord)
## [1] "B" "A"

my.factor3rev <- reorder(my.factor3, -my.vector3) # a simple trick
levels(my.factor3rev)
## [1] "A" "B"
```

In the last statement, using the unary negation operator, which is vectorized, allows us to easily reverse the ordering of the levels, while still using the default function, `mean()`, to summarize the data.

Reordering factor values. It is possible to arrange the values stored in a factor either alphabetically according to the labels of the levels or according to the order of the levels.

```
# gl() keeps order of levels
my.factor4 <- gl(4, 3, labels = c("A", "F", "B", "Z"))
my.factor4
as.integer(my.factor4)
# factor() orders levels alphabetically
my.factor5 <- factor(rep(c("A", "F", "B", "Z"), rep(3,4)))
my.factor5
as.integer(my.factor5)
levels(my.factor5)[as.integer(my.factor5)]
```

We see above that the integer values by which levels in a factor are stored, are equivalent to indices or "subscripts" referencing the vector of labels. Function `sort()` operates on the values' underlying integers and sorts according to the order of the levels while `order()` operates on the values' labels and returns a vector of indices that arrange the values alphabetically.

```
sort(my.factor4)
my.factor4[order(my.factor4)]
my.factor4[order(as.integer(my.factor4))]
```

Run the examples in the chunk above and work out why the results differ.

2.13   Lists

Lists' main difference to other collections is, in R, that they can be heterogeneous. In R, the members of a list can be considered as following a sequence, and accessible through numerical indexes, the same as vectors. However, frequently members of a list are given names, and retrieved (indexed) through these names. Lists are created using function `list()`.

```
a.list <- list(x = 1:6, y = "a", z = c(TRUE, FALSE))
a.list
## $x
## [1] 1 2 3 4 5 6
##
## $y
## [1] "a"
##
## $z
## [1]   TRUE FALSE
```

2.13.1   Member extraction and subsetting

Using double square brackets for indexing a list extracts the element stored in the list, in its original mode. In the example above, `a.list[["x"]]` returns a numeric vector, while `a.list[1]` returns a list containing the numeric vector x. `a.list$x` returns the same value as `a.list[["x"]]`, a numeric vector. `a.list[c(1,3)]` returns a list of length two, while `a.list[[c(1,3)]]` is an error.

```
a.list$x
## [1] 1 2 3 4 5 6

a.list[["x"]]
## [1] 1 2 3 4 5 6

a.list[[1]]
## [1] 1 2 3 4 5 6

a.list["x"]
## $x
## [1] 1 2 3 4 5 6

a.list[1]
## $x
## [1] 1 2 3 4 5 6

a.list[c(1,3)]
## $x
## [1] 1 2 3 4 5 6
##
## $z
## [1]   TRUE FALSE

try(a.list[[c(1,3)]])
## [1] 3
```

> Lists, as usually defined in languages like C, are based on pointers stored at each node, and these pointers chain or link the different member nodes. In such implementations, indexing by position is not possible, or at least requires "walking" down the list, node by node. In R, list members can be accessed through positional indexes. Of course, insertions and deletions in the middle of a list, whatever their implementation, alter (or *invalidate*) any position-based indexes.

To investigate the returned values, function str() (*structure*) is of help, especially when the lists have many members, as it formats lists more compactly than function print().

```
str(a.list)
## List of 3
##  $ x: int [1:6] 1 2 3 4 5 6
##  $ y: chr "a"
##  $ z: logi [1:2] TRUE FALSE
```

2.13.2   Adding and removing list members

In other languages the two most common operations on lists are insertions and deletions. In R, function append() can be used both to append elements at the end of a list and insert elements into the head or any position in the middle of a list.

```
another.list <- append(a.list, list(yy = 1:10, zz = letters[5:1]), 2L)
another.list
## $x
## [1] 1 2 3 4 5 6
##
## $y
## [1] "a"
##
## $yy
##  [1]  1  2  3  4  5  6  7  8  9 10
##
## $zz
## [1] "e" "d" "c" "b" "a"
##
## $z
## [1]  TRUE FALSE
```

To delete a member from a list we assign NULL to it.

```
a.list$y <- NULL
a.list
## $x
## [1] 1 2 3 4 5 6
##
## $z
## [1]  TRUE FALSE
```

Lists can be nested, i.e., lists of lists.

```
a.list <- list("a", "aa", "aaa")
b.list <- list("b", "bb")
nested.list <- list(A = a.list, B = b.list)
str(nested.list)
## List of 2
##  $ A:List of 3
##   ..$ : chr "a"
##   ..$ : chr "aa"
##   ..$ : chr "aaa"
##  $ B:List of 2
##   ..$ : chr "b"
##   ..$ : chr "bb"
```

> ☕ When dealing with deep lists, it is sometimes useful to limit the number of levels of nesting returned by `str()` by means of a `numeric` argument passed to parameter `max.levels`.
>
> ```
> str(nested.list, max.level = 1)
> ## List of 2
> ##  $ A:List of 3
> ##  $ B:List of 2
> ```

2.13.3 Nested lists

A nested list can be constructed within a single statement in which several member lists are created. Here we combine the first three statements in the earlier chunk into a single one.

```
nested.list <- list(A = list("a", "aa", "aaa"), B = list("b", "bb"))
str(nested.list)
## List of 2
##  $ A:List of 3
##   ..$ : chr "a"
##   ..$ : chr "aa"
##   ..$ : chr "aaa"
##  $ B:List of 2
##   ..$ : chr "b"
##   ..$ : chr "bb"
```

> ☕ The logic behind extraction of members of nested lists using indexing is the same as for simple lists, but applied recursively—e.g., `nested.list[[2]]` extracts the second member of the outermost list, which is another list. As, this is a list, its members can be extracted using again the extraction operator: `nested.list[[2]][[1]]`. It is important to remember that these concatenated extraction operations are written so that the leftmost operator is applied to the outermost list. The example given here uses the `[[ ]]` operator, but the same logic applies to `[ ]`.

What do you expect each of the statements below to return? *Before running the code*, predict what value and of which mode each statement will return. You may use implicit or explicit calls to `print()`, or calls to `str()` to visualize the structure of the different objects.

```
nested.list <- list(A = list("a", "aa", "aaa"), B = list("b", "bb"))
str(nested.list)
nested.list[2:1]
nested.list[1]
nested.list[[1]][2]
nested.list[[1]][[2]]
nested.list[2]
nested.list[2][[1]]
```

Sometimes we need to flatten a list, or a nested structure of lists within lists. Function `unlist()` is what should be normally used in such cases.

The list `nested.list` is a nested system of lists, but all the "terminal" members are character strings. In other words, terminal nodes are all of the same mode, allowing the list to be "flattened" into a character vector.

```
c.vec <- unlist(nested.list)
c.vec
##    A1    A2    A3    B1    B2
##   "a"  "aa" "aaa"  "b"  "bb"

is.list(nested.list)
## [1] TRUE

is.list(c.vec)
## [1] FALSE

mode(nested.list)
## [1] "list"

mode(c.vec)
## [1] "character"

names(nested.list)
## [1] "A" "B"

names(c.vec)
## [1] "A1" "A2" "A3" "B1" "B2"
```

The returned value is a vector with named member elements. We use function `str()` to figure out how this vector relates to the original list. The names are based on the names of list elements when available, while numbers are used for anonymous nodes. We can access the members of the vector either through numeric indexes or names.

```
str(c.vec)
##  Named chr [1:5] "a" "aa" "aaa" "b" "bb"
##  - attr(*, "names")= chr [1:5] "A1" "A2" "A3" "B1" ...
```

```
c.vec[2]
##   A2
## "aa"
```

```
c.vec["A2"]
##   A2
## "aa"
```

Function `unname()` can be used to remove names safely—i.e., without risk of altering the mode or class of the object.

```
unname(c.vec)
## [1] "a"   "aa"  "aaa" "b"   "bb"
```

> ![icon] Function `unlist()` has two additional parameters, with default argument values, which we did not modify in the example above. These parameters are `recursive` and `use.names`, both of them expecting a `logical` value as an argument. Modify the statement `c.vec <- unlist(c.list)`, by passing `FALSE` as an argument to these two parameters, in turn, and in each case, study the value returned and how it differs with respect to the one obtained above.

2.14   Data frames

Data frames are a special type of list, in which each element is a vector or a factor of the same length. They are created with function `data.frame()` with a syntax similar to that used for lists—in object-oriented programming we say that data frames are derived from class `list`. As the expectation is equal length, if vectors of different lengths are supplied as arguments, the shorter vector(s) is/are recycled, possibly several times, until the required full length is reached.

```
a.df <- data.frame(x = 1:6, y = "a", z = c(TRUE, FALSE))
a.df
##   x y     z
## 1 1 a  TRUE
## 2 2 a FALSE
## 3 3 a  TRUE
## 4 4 a FALSE
## 5 5 a  TRUE
## 6 6 a FALSE
```

```
str(a.df)
## 'data.frame': 6 obs. of  3 variables:
##  $ x: int  1 2 3 4 5 6
##  $ y: chr  "a" "a" "a" "a" ...
##  $ z: logi  TRUE FALSE TRUE FALSE TRUE FALSE
```

```
class(a.df)
```

```
## [1] "data.frame"

mode(a.df)
## [1] "list"

is.data.frame(a.df)
## [1] TRUE

is.list(a.df)
## [1] TRUE
```

Indexing of data frames is similar to that of the underlying list, but not exactly equivalent. We can index with operator [[]] to extract individual variables, thought of being the columns in a matrix-like list or "worksheet."

```
a.df$x
## [1] 1 2 3 4 5 6

a.df[["x"]]
## [1] 1 2 3 4 5 6

a.df[[1]]
## [1] 1 2 3 4 5 6

class(a.df)
## [1] "data.frame"
```

With function class() we can query the class of an R object (see section 2.8 on page 41). As we saw in the two previous chunks, list and data.frame objects belong to two different classes. However, their relationship is based on a hierarchy of classes. We say that class data.frame is derived from class list. Consequently, data frames inherit the methods and characteristics of lists, as long as they have not been hidden by new ones defined for data frames.

In the same way as with vectors, we can add members to lists and data frames.

```
a.df$x2 <- 6:1
a.df$x3 <- "b"
str(a.df)
## 'data.frame': 6 obs. of  5 variables:
##  $ x : int  1 2 3 4 5 6
##  $ y : chr  "a" "a" "a" "a" ...
##  $ z : logi   TRUE FALSE TRUE FALSE TRUE FALSE
##  $ x2: int  6 5 4 3 2 1
##  $ x3: chr  "b" "b" "b" "b" ...
```

We have added two columns to the data frame, and in the case of column x3 recycling took place. This is where lists and data frames differ substantially in their behavior. In a data frame, although class and mode can be different for different variables (columns), they are required to be vectors or factors of the same length. In the case of lists, there is no such requirement, and recycling never takes place when adding a node. Compare the values returned below for a.ls, to those in the example above for a.df.

```
a.ls <- list(x = 1:6, y = "a", z = c(TRUE, FALSE))
str(a.ls)
## List of 3
##  $ x: int [1:6] 1 2 3 4 5 6
##  $ y: chr "a"
##  $ z: logi [1:2] TRUE FALSE

a.ls$x2 <- 6:1
a.ls$x3 <- "b"
str(a.ls)
## List of 5
##  $ x : int [1:6] 1 2 3 4 5 6
##  $ y : chr "a"
##  $ z : logi [1:2] TRUE FALSE
##  $ x2: int [1:6] 6 5 4 3 2 1
##  $ x3: chr "b"
```

Data frames are extremely important to anyone analyzing or plotting data using R. One can think of data frames as tightly structured work-sheets, or as lists. As you may have guessed from the examples earlier in this section, there are several different ways of accessing columns, rows, and individual observations stored in a data frame. The columns can be treated as members in a list, and can be accessed both by name or index (position). When accessed by name, using $ or double square brackets, a single column is returned as a vector or factor. In contrast to lists, data frames are always "rectangular" and for this reason the values stored can also be accessed in a way similar to how elements in a matrix are accessed, using two indexes. As we saw for vectors, indexes can be vectors of integer numbers or vectors of logical values. For columns they can, in addition, be vectors of character strings matching the names of the columns. When using indexes it is extremely important to remember that the indexes are always given **row first**.

☕ Indexing of data frames can in all cases be done as if they were lists, which is preferable, as it ensures compatibility with regular R lists and with newer implementations of data-frame-like structures like those defined in package 'tibble'. Using this approach, extracting two values from the second and third positions in the first column of `a.df` is done as follows, using numerical indexes.

```
a.df[[1]][2:3]
## [1] 2 3
```

Or using the column name.

```
a.df[["x"]][2:3]
## [1] 2 3
```

The less portable, matrix-like indexing is done as follows, with the first index indicating rows and the second one indicating columns. This notation allows simultaneous extraction from multiple columns, which is not possible with lists. The value returned is a "smaller" data frame.

```
a.df[2:3, 1:2]
##   x y
## 2 2 a
## 3 3 a
```

If the length of the column indexing vector is one, the returned value is a vector, which is not consistent with the previous example which returned a data frame. This is not only surprising in everyday use, but can be the source of bugs when coding algorithms in which the length of the second index vector cannot be guaranteed to be always more than one.

```
a.df[2:3, 1]
## [1] 2 3
```

In contrast, indexing of tibbles—defined in package 'tibble'—is always consistent, never returning a vector, even when the vector used to extract columns is of length one (see section 6.4.2 on page 182 for details on the differences between data frames and tibbles).

```
# first column, a.df[[1]] preferred
a.df[ , 1]
## [1] 1 2 3 4 5 6

# first column, a.df[["x"]] or a.df$x preferred
a.df[ , "x"]
## [1] 1 2 3 4 5 6

# first row
a.df[1, ]
##   x y    z x2 x3
## 1 1 a TRUE  6  b

# first two rows of the third and fourth columns
a.df[1:2, c(FALSE, FALSE, TRUE, TRUE, FALSE)]
##       z x2
## 1  TRUE  6
## 2 FALSE  5

# the rows for which z is true
a.df[a.df$z , ]
##   x y    z x2 x3
## 1 1 a TRUE  6  b
## 3 3 a TRUE  4  b
## 5 5 a TRUE  2  b

# the rows for which x > 3 keeping all columns except the third one
a.df[a.df$x > 3, -3]
##   x y x2 x3
## 4 4 a  3  b
## 5 5 a  2  b
## 6 6 a  1  b
```

As explained earlier for vectors (see section 2.10 on page 45), indexing can be

present both on the right-hand side and left-hand side of an assignment. The next
few examples do assignments to "cells" of a.df, either to one whole column, or
individual values. The last statement in the chunk below copies a number from
one location to another by using indexing of the same data frame both on the
right side and left side of the assignment.

```
a.df[1, 1] <- 99
a.df
##     x y      z x2 x3
## 1 99 a  TRUE  6  b
## 2  2 a FALSE  5  b
## 3  3 a  TRUE  4  b
## 4  4 a FALSE  3  b
## 5  5 a  TRUE  2  b
## 6  6 a FALSE  1  b

a.df[ , 1] <- -99
a.df
##      x y      z x2 x3
## 1 -99 a  TRUE  6  b
## 2 -99 a FALSE  5  b
## 3 -99 a  TRUE  4  b
## 4 -99 a FALSE  3  b
## 5 -99 a  TRUE  2  b
## 6 -99 a FALSE  1  b

a.df[["x"]] <- 123
a.df
##      x y      z x2 x3
## 1 123 a  TRUE  6  b
## 2 123 a FALSE  5  b
## 3 123 a  TRUE  4  b
## 4 123 a FALSE  3  b
## 5 123 a  TRUE  2  b
## 6 123 a FALSE  1  b

a.df[1, 1] <- a.df[6, 4]
a.df
##      x y      z x2 x3
## 1   1 a  TRUE  6  b
## 2 123 a FALSE  5  b
## 3 123 a  TRUE  4  b
## 4 123 a FALSE  3  b
## 5 123 a  TRUE  2  b
## 6 123 a FALSE  1  b
```

> ⚠ We mentioned above that indexing by name can be done either with dou-
> ble square brackets, [[]], or with $. In the first case the name of the variable
> or column is given as a character string, enclosed in quotation marks, or as a
> variable with mode character. When using $, the name is entered as a constant,
> without quotation marks, and cannot be a variable.

```
x.list <- list(abcd = 123, xyzw = 789)
x.list[["abcd"]]
## [1] 123

a.var <- "abcd"
x.list[[a.var]]
## [1] 123


x.list$abcd
## [1] 123

x.list$ab
## [1] 123

x.list$a
## [1] 123
```

Both in the case of lists and data frames, when using double square brackets, an exact match is required between the name in the object and the name used for indexing. In contrast, with $, any unambiguous partial match will be accepted. For interactive use, partial matching is helpful in reducing typing. However, in scripts, and especially R code in packages, it is best to avoid the use of $ as partial matching to a wrong variable present at a later time, e.g., when someone else revises the script, can lead to very difficult-to-diagnose errors. In addition, as $ is implemented by first attempting a match to the name and then calling [[]], using $ for indexing can result in slightly slower performance compared to using [[]]. It is possible to set an R option so that partial matching triggers a warning, which can be very useful when debugging.

2.14.1 Operating within data frames

When the names of data frames are long, complex conditions become awkward to write using indexing—i.e., subscripts. In such cases subset() is handy because evaluation is done in the "environment" of the data frame, i.e., the names of the columns are recognized if entered directly when writing the condition. Function subset() "filters" rows, usually corresponding to observations or experimental units. The condition is computed for each row, and if it returns TRUE, the row is included in the returned data frame, and excluded if FALSE.

```
a.df <- data.frame(x = 1:6, y = "a", z = c(TRUE, FALSE))
subset(a.df, x > 3)
##   x y     z
## 4 4 a FALSE
## 5 5 a  TRUE
## 6 6 a FALSE
```

What is the behavior of subset() when the condition is NA? Find the answer by writing code to test this, for a case where tests for different rows return NA, TRUE and FALSE.

When calling functions that return a vector, data frame, or other structure, the square brackets can be appended to the rightmost parenthesis of the function call, in the same way as to the name of a variable holding the same data.

```
subset(a.df, x > 3)[ , -3]
##   x y
## 4 4 a
## 5 5 a
## 6 6 a

subset(a.df, x > 3)$x
## [1] 4 5 6
```

None of the examples in the last three code chunks alter the original data frame a.df. We can store the returned value using a new name if we want to preserve a.df unchanged, or we can assign the result to a.df, deleting in the process, the previously stored value.

> ⚠ In the example above, the names in the expression passed as the second argument to subset() are first searched within ad.df but if not found, searched in the environment. There being no variable A in the data frame a.df, vector A from the environment is silently used in the expression resulting in a returned data frame with no rows.
>
> ```
> A <- 1
> subset(a.df, A > 3)
> ## [1] x y z
> ## <0 rows> (or 0-length row.names)
> ```
>
> The use of subset() is convenient, but more prone to result in bugs compared to directly using the extraction operator []. This same "cost" to achieving convenience applies to functions like attach() and with() described below. The longer time that a script is expected to be used, adapted and reused, the more careful we should be when using any of these functions. An alternative way of avoiding excessive verbosity is to keep the names of data frames short.

A frequently used way of deleting a column by name from a data frame is to assign NULL to it—i.e., in the same way as members are deleted from lists. This approach modifies a.df in place.

```
aa.df <- a.df
colnames(aa.df)
## [1] "x" "y" "z"

aa.df[["y"]] <- NULL
colnames(aa.df)
## [1] "x" "z"
```

Alternatively, we can use negative indexing to remove columns from a copy of a data frame. In this example we remove a single column. As base R does not support negative indexing by name, we need to find the numerical index of the column to delete.

```
a.df[ , -which(colnames(a.df) == "y")]
##   x    z
## 1 1  TRUE
## 2 2 FALSE
## 3 3  TRUE
## 4 4 FALSE
## 5 5  TRUE
## 6 6 FALSE
```

Instead of using the equality test, we can use the operator %in% or function grepl() to delete multiple columns in a single statement.

In the previous code chunk we deleted the last column of the data frame a.df. Here is an esoteric trick for you to first untangle how it changes the positions of columns and row, and then for you to think how and why it can be useful to use indexing with the extraction operator [ ] on both sides of the assignment operator <-.

```
a.df[1:6, c(1,3)] <- a.df[6:1, c(3,1)]
a.df
```

Although in this last example we used numeric indexes to make it more interesting, in practice, especially in scripts or other code that will be reused, do use column or member names instead of positional indexes whenever possible. This makes code much more reliable, as changes elsewhere in the script could alter the order of columns and *invalidate* numerical indexes. In addition, using meaningful names makes programmers' intentions easier to understand.

It is sometimes inconvenient to have to pre-pend the name of a *container* such as a list or data frame to the name of each member variable being accessed. There are functions in R that allow us to change where R looks for the names of objects we include in a code statement. Here I describe the use of attach() and its matching detach(), and with() and within() to access members of a data frame. They can be used as well with lists and classes derived from list.

As we can see below, when using a rather long name for a data frame, enter-

ing a simple calculation can easily result in a long and difficult to read statement. (Method head() is used here to limit the displayed value to the first two
rows—head() is described in section 2.17 on page 81.)

```
my_data_frame.df <- data.frame(A = 1:10, B = 3)
my_data_frame.df$C <-
  (my_data_frame.df$A + my_data_frame.df$B) / my_data_frame.df$A
head(my_data_frame.df, 2)
##   A B   C
## 1 1 3 4.0
## 2 2 3 2.5
```

Using attach() we can alter how R looks up names and consequently simplify the statement. With detach() we can restore the original state. It is important to remember that here we can only simplify the right-hand side of the
assignment, while the "destination" of the result of the computation still needs
to be fully specified on the left-hand side of the assignment operator. We include below only one statement between attach() and detach() but multiple
statements are allowed. Furthermore, if variables with the same name as the
columns exist in the search path, these will take precedence, something that
can result in bugs or crashes, or as seen below, a message warns that variable
A from the global environment will be used instead of column A of the attached
my_data_frame.df. The returned value is, of course, not the desired one.

```
my_data_frame.df$C <- NULL
attach(my_data_frame.df)

## The following object is masked _by_ .GlobalEnv:
##
##     A

my_data_frame.df$C <- (A + B) / A
detach(my_data_frame.df)
head(my_data_frame.df, 2)
##   A B C
## 1 1 3 4
## 2 2 3 4
```

In the case of with() only one, possibly compound code statement is affected and this statement is passed as an argument. As before, we need to
fully specify the left-hand side of the assignment. The value returned is the
one returned by the statement passed as an argument, in the case of compound
statements, the value returned by the last contained simple code statement to
be executed. Consequently, if the intent is to modify the container, assignment
to an individual member variable (column in this case) is required. In contrast
to the behavior of attach(), In this case, column A of my_data_frame.df takes
precedence, and the returned value is the expected one.

```
my_data_frame.df$C <- NULL
my_data_frame.df$C <- with(my_data_frame.df, (A + B) / A)
head(my_data_frame.df, 2)
##   A B   C
## 1 1 3 4.0
## 2 2 3 2.5
```

In the case of within(), assignments in the argument to its second parameter affect the object returned, which is a copy of the container (In this case, a
whole data frame), which still needs to be saved through assignment. Here the
intention is to modify it, so we assign it back to the same name, but it could
have been assigned to a different name so as not to overwrite the original data
frame.

```
my_data_frame.df$C <- NULL
my_data_frame.df <- within(my_data_frame.df,  C <- (A + B) / A)
head(my_data_frame.df, 2)
##   A B   C
## 1 1 3 4.0
## 2 2 3 2.5
```

In the example above, within() makes little difference compared to using
with() with respect to the amount of typing or clarity, but with multiple member variables being operated upon, as shown below, within() has an advantage
resulting in more concise and easier to understand code.

```
my_data_frame.df$C <- NULL
my_data_frame.df <- within(my_data_frame.df,
                           {C <- (A + B) / A
                            D <- A * B
                            E <- A / B + 1}
                           )
head(my_data_frame.df, 2)
##   A B        E D   C
## 1 1 3 1.333333 3 4.0
## 2 2 3 1.666667 6 2.5
```

Use of attach() and detach(), which function as a pair of ON and OFF
switches, can result in an undesired after-effect on name lookup if the script
terminates after attach() is executed but before detach() is called, as cleanup
is not automatic. In contrast, with() and within(), being self-contained, guarantee that cleanup takes place. Consequently, the usual recommendation is to
give preference to the use of with() and within() over attach() and detach().
Use of these functions not only saves typing but also makes code more readable.

2.14.2 Re-arranging columns and rows

The most direct way of changing the order of columns and/or rows in data frames
(and matrices and arrays) is to use subscripting as described above. Once we know

the original position and target position we can use numerical indexes on both right-hand side and left-hand side of an assignment.

> ⚠ When using the extraction operator [] on both the left-hand-side and right-hand-side to swap columns, the vectors or factors are swapped, while the names of the columns are not! The same applies to row names, which makes storing important information in them inconvenient and error prone.

To retain the correct naming after the column swap, we need to separately swap the names of the columns.

```
my_data_frame.df <- data.frame(A = 1:10, B = 3)
head(my_data_frame.df, 2)
##   A B
## 1 1 3
## 2 2 3

my_data_frame.df[ , 1:2] <- my_data_frame.df[ , 2:1]
head(my_data_frame.df, 2)
##   A B
## 1 3 1
## 2 3 2

colnames(my_data_frame.df)[1:2] <- colnames(my_data_frame.df)[2:1]
head(my_data_frame.df, 2)
##   B A
## 1 3 1
## 2 3 2
```

Taking into account that order() returns the indexes needed to sort a vector (see page 49), we can use order() to generate the indexes needed to sort either columns or rows of a data frame. When we want to sort rows, the argument to order() is usually a column of the data frame being arranged. However, any vector of suitable length, including the result of applying a function to one or more columns, can be passed as an argument to order().

> 📇 The first task to be completed is to sort a data frame based on the values in one column, using indexing and order(). Create a new data frame and with three numeric columns with three different haphazard sequences of values. Call these columns A, B and C. 1) Sort the rows of the data frame so that the values in A are in decreasing order. 2) Sort the rows of the data frame according to increasing values of the sum of A and B without adding a new column to the data frame or storing the vector of sums in a variable. In other words, do the sorting based on sums calculated on the fly.

> Repeat the tasks in the playground immediately above but using fac-

tors instead of numeric vectors as columns in the data frame. Hint: revisit the exercise on page 61 were the use of `order()` on factors is described.

2.15 Attributes of R objects

R objects can have attributes. Attributes are normally used to store ancillary data. They are used by R itself to store things like column names in data frames and labels of factor levels. All these attributes are visible to user code, and user code can read and write objects' attributes. Attribute `"comment"` is meant to be set by users—e.g., to store metadata together with data.

```
a.df <- data.frame(x = 1:6, y = "a", z = c(TRUE, FALSE))
comment(a.df)
## NULL

comment(a.df) <- "this is stored as a comment"
comment(a.df)
## [1] "this is stored as a comment"
```

Methods like `names()`, `dim()` or `levels()` return values retrieved from attributes stored in R objects, and methods like `names()<-`, `dim()<-` or `levels()<-` set (or unset with `NULL`) the value of the respective attributes. Specific query and set methods do not exist for all attributes. Methods `attr()`, `attr()<-` and `attributes()` can be used with any attribute. In addition, method `str()` displays all components of R objects including their attributes.

```
names(a.df)
## [1] "x" "y" "z"

names(a.df) <- toupper(names(a.df))
names(a.df)
## [1] "X" "Y" "Z"

attr(a.df, "names") # same as previous line
## [1] "X" "Y" "Z"

attr(a.df, "my.attribute") <- "this is stored in my attribute"
attributes(a.df)
## $names
## [1] "X" "Y" "Z"
##
## $class
## [1] "data.frame"
##
## $row.names
## [1] 1 2 3 4 5 6
##
## $comment
## [1] "this is stored as a comment"
##
## $my.attribute
## [1] "this is stored in my attribute"
```

⚠️ There is no restriction to the creation, setting, resetting and reading of attributes, but not all methods and operators that can be used to modify objects will preserve non-standard attributes. So, using private attributes is a double-edged sword that usually is worthwhile considering only when designing a new class together with the corresponding methods for it. A good example of extensive use of class-specific attributes are the values returned by model fitting functions like lm() (see section 4.6 on page 127).

Even the class of S3 objects is stored as an attribute that is accessible as any other attribute—this is in contrast to the mode and atomic class of an object. Object-oriented programming in R in explained in section 5.4 on page 172.

```
numbers <- 1:10
class(numbers)
## [1] "integer"

attributes(numbers)
## NULL


a.factor <- factor(numbers)
class(a.factor)
## [1] "factor"

attributes(a.factor)
## $levels
##  [1] "1"  "2"  "3"  "4"  "5"  "6"  "7"  "8"  "9"  "10"
##
## $class
## [1] "factor"
```

2.16   Saving and loading data

2.16.1   Data sets in R and packages

To be able to present more meaningful examples, we need some real data. Here we use cars, one of the many data sets included in base R. Function data() is used to load data objects that are included in R or contained in packages. It is also possible to import data saved in files with *foreign* formats, defined by other software or commonly used for data exchange. Package 'foreign', included in the R distribution, as well as contributed packages make available functions capable of reading and decoding various foreign formats. How to read or import "foreign" data is discussed in R documentation in *R Data Import/Export*, and in this book, in chapter 8 starting on page 293. It is also good to keep in mind that in R, URLs (Uniform Resource Locators) are accepted as arguments to the file or path parameter of many functions (see section 8.12 starting on page 322).

In the next example we load data included in R as R objects by calling function `data()`. The loaded R object `cars` is a data frame.

```
data(cars)
```

Once we have a data set available, the first step is usually to explore it, and we will do this with `cars` in section 2.17 on page 81.

2.16.2 .rda files

By default, at the end of a session, the current workspace containing the results of your work is saved into a file called .RData. In addition to saving the whole workspace, it is possible to save one or more R objects present in the workspace to disk using the same file format (with file name tag .rda or .Rda). One or more objects, belonging to any mode or class can be saved into a single file using function `save()`. Reading the file restores all the saved objects into the current workspace with their original names. These files are portable across most R versions—i.e., old formats can be read and written by newer versions of R, although the newer, default format may be not readable with earlier R versions. Whether compression is used, and whether the "binary" data is encoded into ASCII characters, allowing maximum portability at the expense of increased size can be controlled by passing suitable arguments to `save()`.

We create a data frame object and then save it to a file.

```
my.df <- data.frame(x = 1:5, y = 5:1)
my.df
##   x y
## 1 1 5
## 2 2 4
## 3 3 3
## 4 4 2
## 5 5 1

save(my.df, file = "my-df.rda")
```

We delete the data frame object and confirm that it is no longer present in the workspace.

```
rm(my.df)
ls(pattern = "my.df")
## character(0)
```

We read the file we earlier saved to restore the object.

```
load(file = "my-df.rda")
ls(pattern = "my.df")
## [1] "my.df"

my.df
##   x y
## 1 1 5
## 2 2 4
## 3 3 3
## 4 4 2
## 5 5 1
```

The default format used is binary and compressed, which results in smaller files.

In the example above, only one object was saved, but one can simply give the names of additional objects as arguments. Just try saving more than one data frame to the same file. Then the data frames plus a few vectors. After creating each file, clear the workspace and then restore from the file the objects you saved.

Sometimes it is easier to supply the names of the objects to be saved as a vector of character strings passed as an argument to parameter `list`. One case is when wanting to save a group of objects based on their names. We can use `ls()` to list the names of objects matching a simple `pattern` or a complex regular expression. The example below does this in two steps, first saving a character vector with the names of the objects matching a pattern, and then using this saved vector as an argument to `save`'s `list` parameter.

```r
objcts <- ls(pattern = "*.df")
save(list = objcts, file = "my-df1.rda")
```

The two statements above can be combined into a single statement by nesting the function calls.

```r
save(list = ls(pattern = "*.df"), file = "my-df1.rda")
```

Practice using different patterns with `ls()`. You do not need to save the objects to a file. Just have a look at the list of object names returned.

As a coda, we show how to clean up by deleting the two files we created. Function `unlink()` can be used to delete any files for which the user has enough rights.

```r
unlink(c("my-df.rda", "my-df1.rda"))
```

2.16.3 .rds files

The RDS format can be used to save individual objects instead of multiple objects (usually using file name tag `.rds`). They are read and saved with functions `readRDS()` and `saveRDS()`, respectively. When RDS files are read, different from when RDA files are loaded, we need to assign the object read to a possibly different name for it to added to the search pass. Of course, it is also possible to use the returned object as an argument to a function or in an expression without saving it to a variable.

```
saveRDS(my.df, "my-df.rds")
```

If we read the file, by default the read R object will be printed at the console.

```
readRDS("my-df.rds")
##   x y
## 1 1 5
## 2 2 4
## 3 3 3
## 4 4 2
## 5 5 1
```

In the next example we assign the read object to a different name, and check that the object read is identical to the one saved.

```
my_read.df <- readRDS("my-df.rds")
identical(my.df, my_read.df)
## [1] TRUE
```

As above, we clean up by deleting the file.

```
unlink("my-df.rds")
```

2.17   Looking at data

There are several functions in R that let us obtain different views into objects. Function print() is useful for small data sets, or objects. Especially in the case of large data frames, we need to explore them step by step. In the case of named components, we can obtain their names with colnames(), rownames(), and names(). If a data frame contains many rows of observations, head() and tail() allow us to easily restrict the number of rows printed. Functions nrow() and ncol() return the number of rows and columns in the data frame (also applicable to matrices but not to lists or vectors where we use length()). As mentioned earlier, function str() concisely displays the structure of R objects.

```
class(cars)
## [1] "data.frame"

nrow(cars)
## [1] 50

ncol(cars)
## [1] 2

names(cars)
## [1] "speed" "dist"

head(cars)
```

```
##   speed dist
## 1     4    2
## 2     4   10
## 3     7    4
## 4     7   22
## 5     8   16
## 6     9   10

tail(cars)
##    speed dist
## 45    23   54
## 46    24   70
## 47    24   92
## 48    24   93
## 49    24  120
## 50    25   85

str(cars)
## 'data.frame': 50 obs. of  2 variables:
##  $ speed: num  4 4 7 7 8 9 10 10 10 11 ...
##  $ dist : num  2 10 4 22 16 10 18 26 34 17 ...
```

> Look up the help pages for `head()` and `tail()`, and edit the code above to print only the first two lines, or only the last three lines of `cars`, respectively.

The different columns of a data frame can be factors or vectors of various modes (e.g., numeric, logical, character, etc.) (see section 2.14 on page 66). To explore the mode of the columns of `cars`, we can use an *apply* function. In the present case, we want to apply function `class()` to each column of the data frame `cars`. (Apply functions are described in section 3.4 on page 108.)

```
sapply(X = cars, FUN = class)
##     speed      dist
## "numeric" "numeric"
```

The statement above returns a vector of character strings, with the mode of each column. Each element of the vector is named according to the name of the corresponding "column" in the data frame. For this same statement to be used with any other data frame or list, we need only to substitute the name of the object, the argument to the first parameter called x, to the one of current interest.

> Data set `airquality` contains data from air quality measurements in New York, and, being included in the R distribution, can be loaded with `data(airquality)`. Load it, and repeat the steps above, to learn what variables (columns) it contains, their classes, the number of rows, etc.

Function `summary()` can be used to obtain a summary from objects of most R classes, including data frames. We can also use `sapply()`, `lapply()` or `vapply()` to apply any suitable function to individual columns.

```
summary(cars)
##      speed           dist
## Min.   : 4.0   Min.   :  2.00
## 1st Qu.:12.0   1st Qu.: 26.00
## Median :15.0   Median : 36.00
## Mean   :15.4   Mean   : 42.98
## 3rd Qu.:19.0   3rd Qu.: 56.00
## Max.   :25.0   Max.   :120.00

sapply(cars, range)
##      speed dist
## [1,]     4    2
## [2,]    25  120
```

Obtain the summary of `airquality` with function `summary()`, but in addition, write code with an *apply* function to count the number of non-missing values in each column. Hint: using `sum()` on a `logical` vector returns the count of `TRUE` values as `TRUE`, and `FALSE` are transparently converted into `numeric` 1 and 0, respectively, when `logical` values are used in arithmetic expressions.

2.18 Plotting

The base-R generic method `plot()` can be used to plot different data. It is a generic method that has specializations suitable for different kinds of objects (see section 5.4 on page 172 for a brief introduction to objects, classes and methods). In this section we only very briefly demonstrate the use of the most common base-R graphics functions. They are well described in the book *R Graphics* (Murrell 2019). We will not describe the Lattice (based on S's Trellis) approach to plotting (Sarkar 2008). Instead we describe in detail the use of the *grammar of graphics* and plotting with package 'ggplot2' in chapter 7 starting on page 203.

It is possible to pass two variables (here columns from a data frame) directly as arguments to the x and y parameters of `plot()`.

```
plot(x = cars$speed, y = cars$dist)
```

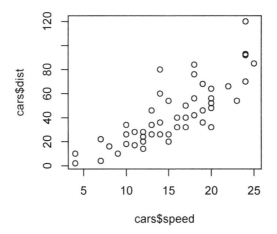

It is also possible, and usually more convenient, to use a *formula* to specify the variables to be plotted on the x and y axes, passing additionally as an argument to parameter `data` the name of the data frame containing these variables. The formula `dist ~ speed`, is read as `dist` explained by `speed`—i.e., `dist` is mapped to the y-axis as the dependent variable and `speed` to the x-axis as the independent variable.

```
plot(dist ~ speed, data = cars)
```

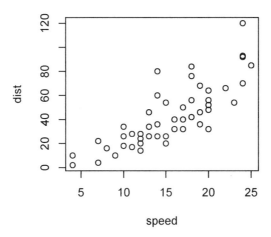

Within R there exist different specializations, or "flavors," of method `plot()` that become active depending on the class of the variables passed as arguments: passing two numerical variables results in a scatter plot as seen above. In contrast passing one factor and one numeric variable to `plot()` results in a box-and-whiskers plot being produced. To exemplify this we need to use a different data set, here `chickwts` as `cars` does not contain any factors. Use `help("chickwts")` to learn more about this data set, also included in R.

```
plot(weight ~ feed, data = chickwts)
```

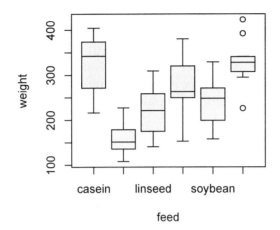

Method `plot()` and variants defined in R, when used for plotting return their graphical output to a *graphical output device*. In R, graphical devices are very frequently not real physical devices like a printer, but virtual devices implemented fully in software that translate the plotting commands into a specific graphical file format. Several different graphical devices are available in R and they differ in the kind of output they produce: raster files (e.g., TIFF, PNG and JPEG formats), vector graphics files (e.g., SVG, EPS and PDF) or output to a physical device like a window in the screen of a computer. Additional devices are available through contributed R packages.

Devices follow the paradigm of ON and OFF switches. Some devices producing a file as output, save this output only when the device is closed. When opening a device the user supplies additional information. For the PDF device that produces output in a vector-graphics format, width and height of the output are specified in *inches*. A default file name is used unless we pass a `character` string as an argument to parameter `file`.

```
pdf(file = "output/my-file.pdf", width = 6, height = 5, onefile = TRUE)
plot(dist ~ speed, data = cars)
plot(weight ~ feed, data = chickwts)
dev.off()
## cairo_pdf
##         2
```

Raster devices return bitmaps and `width` and `height` are specified in *pixels*.

```
png(file = "output/my-file.png", width = 600, height = 500)
plot(weight ~ feed, data = chickwts)
dev.off()
## cairo_pdf
##         2
```

When R is used interactively, a device to output the graphical output to a display device is opened automatically. The name of the device may depend on the operating system used (e.g., MS-Windows or Linux) or an IDE—e.g., RStudio defines its own graphic device for output to the Plots pane of its user interface.

⚠ This approach of direct output to a device, and addition of plot components as shown below directly on the output device itself, is not the only approach available. As we will see in chapter 7 starting on page 203, an alternative approach is to build a *plot object* as a list of member components that is later rendered as a whole on a graphical device by calling `print()` once.

```
png(file = "output/my-file.png", width = 600, height = 500)
plot(dist ~ speed, data = cars)
text(x = 10, y = 110, labels = "some texts to be added")
dev.off()
## cairo_pdf
##         2
```

2.19   Further reading

For further reading on the aspects of R discussed in the current chapter, I suggest the books *R Programming for Data Science* (Peng) and *The Art of R Programming: A Tour of Statistical Software Design* (Matloff).

3

The R language: "Paragraphs" and "essays"

An R script is simply a text file containing (almost) the same commands that you would enter on the command line of R.

Jim Lemon
Kickstarting R

3.1 Aims of this chapter

For those who have mainly used graphical user interfaces, understanding why and when scripts can help in communicating a certain data analysis protocol can be revelatory. As soon as a data analysis stops being trivial, describing the steps followed through a system of menus and dialogue boxes becomes extremely tedious.

Moreover, graphical user interfaces tend to be difficult to extend or improve in a way that keeps step-by-step instructions valid across program versions and operating systems.

Many times, exactly the same sequence of commands needs to be applied to different data sets, and scripts make both implementation and validation of such a requirement easy.

In this chapter, I will walk you through the use of R scripts, starting from an extremely simple script.

3.2 Writing scripts

In R language, the closest match to a natural language essay is a script. A script is built from multiple interconnected code statements needed to complete a given task. Simple statements can be combined into compound statements, which are the equivalent of natural language paragraphs. Scripts can vary from simple scripts containing only a few code statements, to complex scripts containing hundreds of code statements. In the rest of the present section I discuss how to write readable and reliable scripts and how to use them.

3.2.1   What is a script?

A *script* is a text file that contains (almost) the same commands that you would type at the console prompt. A true script is not, for example, an MS-Word file where you have pasted or typed some R commands. A script file has the following characteristics.

- The script is a text file.

- The file contains valid R statements (including comments) and nothing else.

- Comments start at a # and end at the end of the line.

- The R statements are in the file in the order that they must be executed.

- R scripts have file names ending in .r or .R.

It is good practice to write scripts so that they are self-contained. To make a script self-contained, one must include calls to `library()` to load the packages used, load or import data from files, perform the data analysis and display and/or save the results of the analysis. Such scripts can be used to apply the same analysis algorithm to other data and/or to reproduce the same analysis at a later time. Such scripts document all steps used for the analysis.

3.2.2   How do we use a script?

A script can be "sourced" using function `source()`. If we have a text file called `my.first.script.r` containing the following text:

```
# this is my first R script
print(3 + 4)
```

and then source this file:

```
source("my.first.script.r")
## [1] 7
```

The results of executing the statements contained in the file will appear in the console. The commands themselves are not shown (by default the sourced file is not *echoed* to the console) and the results will not be printed unless you include explicit `print()` commands in the script. This applies in many cases also to plots— e.g., a figure created with `ggplot()` needs to be printed if we want it to be included in the output when the script is run. Adding a redundant `print()` is harmless.

From within RStudio, if you have an R script open in the editor, there will be a "source" icon visible with an attached drop-down menu from which you can choose "Source" as described above, or "Source with echo," or "Source as local job" for the script in the currently active editor tab.

When a script is *sourced*, the output can be saved to a text file instead of being shown in the console. It is also easy to call R with the R script file as an argument directly at the operating system shell or command-interpreter prompt—and obviously also from shell scripts. The next two chunks show commands entered at the OS shell command prompt rather than at the R command prompt.

```
> RScript my.first.script.r
```

You can open an operating system's *shell* from the Tools menu in RStudio, to run this command. The output will be printed to the shell console. If you would like to save the output to a file, use redirection using the operating system's syntax.

```
> RScript my.first.script.r > my.output.txt
```

Sourcing is very useful when the script is ready, however, while developing a script, or sometimes when testing things, one usually wants to run (or *execute*) one or a few statements at a time. This can be done using the "run" button[1] after either positioning the cursor in the line to be executed, or selecting the text that one would like to run (the selected text can be part of a line, a whole line, or a group of lines, as long as it is syntactically valid). The key-shortcut Ctrl-Enter is equivalent to pressing the "run" button in RStudio.

3.2.3 How to write a script

As with any type of writing, different approaches may be preferred by different R users. In general, the approach used, or mix of approaches, will also depend on how confident you are that the statements will work as expected—you already know the best approach vs. you are exploring different alternatives.

If one is very familiar with similar problems One would just create a new text file and write the whole thing in the editor, and then test it. This is rather unusual.

If one is moderately familiar with the problem One would write the script as above, but testing it, step by step, as one is writing it. This is usually what I do.

If one is mostly playing around Then if one is using RStudio, one can type statements at the console prompt. As you should know by now, everything you run at the console is saved to the "History." In RStudio, the History is displayed in its own pane, and in this pane one can select any previous statement(s) and by clicking on a single icon, copy and paste them to either the R console prompt, or the cursor position in the editor pane. In this way one can build a script by copying and pasting from the history to your script file, the bits that have worked as you wanted.

---

[1] If you use a different IDE or editor with an R mode, the details will vary, but a run command will be usually available.

> 📖 By now you should be familiar enough with R to be able to write your own script.
>
> 1. Create a new R script (in RStudio, from the File menu, "+" icon, or by typing "Ctrl + Shift + N").
> 2. Save the file as `my.second.script.r`.
> 3. Use the editor pane in RStudio to type some R commands and comments.
> 4. *Run* individual commands.
> 5. *Source* the whole file.

3.2.4 The need to be understandable to people

When you write a script, it is either because you want to document what you have done or you want re-use the script at a later time. In either case, the script itself although still meaningful for the computer, could become very obscure to you, and even more to someone seeing it for the first time. This must be avoided by spending time and effort on the writing style.

How does one achieve an understandable script or program?

• Avoid the unusual. People using a certain programming language tend to use some implicit or explicit rules of style—style includes *indentation* of statements, *capitalization* of variable and function names. As a minimum try to be consistent with yourself.

• Use meaningful names for variables, and any other object. What is meaningful depends on the context. Depending on common use, a single letter may be more meaningful than a long word. However self-explanatory names are usually better: e.g., using `n.rows` and `n.cols` is much clearer than using `n1` and `n2` when dealing with a matrix of data. Probably `number.of.rows` and `number.of.columns` would make the script verbose, and take longer to type without gaining anything in return.

• How to make the words visible in names: traditionally in R one would use dots to separate the words and use only lower case. Some years ago, it became possible to use underscores. The use of underscores is quite common nowadays because in some contexts it is "safer", as in some situations a dot may have a special meaning. What we call "camel case" is only infrequently used in R programming but is common in other languages like Pascal. An example of camel case is `NumCols`.

> 📖 Here is an example of bad style in a script. Read *Google's R Style Guide* (`https://google.github.io/styleguide/Rguide.xml`), and edit the code in the chunk below so that it becomes easier to read.

```
a <- 2 # height
b <- 4 # length
c <-
    a *
b
c -> variable
    print(
"area: ", variable
)
```

The points discussed above already help a lot. However, one can go further in achieving the goal of human readability by interspersing explanations and code "chunks" and using all the facilities of typesetting, even of formatted maths formulas and equations, within the listing of the script. Furthermore, by including the results of the calculations and the code itself in a typeset report built automatically, we can ensure that the results are indeed the result of running the code shown. This greatly contributes to data analysis reproducibility, which is becoming a widespread requirement for any data analysis both in academia and in industry. It is possible not only to typeset whole books like this one, but also whole data-based web sites with these tools.

In the realm of programming, this approach is called literate programming and was first proposed by Donald Knuth (Knuth 1984) through his WEB system. In the case of R programming, the first support of literate programming was through 'Sweave', which has been mostly superseded by 'knitr' (Xie 2013). This package supports the use of Markdown or LaTeX (Lamport 1994) as the markup language for the textual contents and also formats and adds syntax highlighting to code chunks. Rmarkdown is an extension to Markdown that makes it easier to include R code in documents (see `http://rmarkdown.rstudio.com/`). It is the basis of R packages that support typesetting large and complex documents ('bookdown'), web sites ('blogdown'), package vignettes ('pkgdown') and slides for presentations (Xie 2016; Xie et al. 2018). The use of 'knitr' is very well integrated into the RStudio IDE.

This is not strictly an R programming subject, as it concerns programming in any language. On the other hand, this is an incredibly important skill to learn, but well described in other books and web sites cited in the previous paragraph. This whole book, including figures, has been generated using 'knitr' and the source code for the book is available through Bitbucket at `https://bitbucket.org/aphalo/learnr-book`.

3.2.5   Debugging scripts

The use of the word *bug* to describe a problem in computer hardware and software started in 1946 when a real bug, more precisely a moth, got between the contacts of a relay in an electromechanical computer causing it to malfunction and Grace Hooper described the first computer *bug*. The use of the term bug in engineering predates the use in computer science, and consequently, the first use of bug in computing caught on easily because it represented an earlier-used metaphor becoming real.

A suitable quotation from a letter written by Thomas Alva Edison 1878 (as given by Hughes 2004):

> It has been just so in all of my inventions. The first step is an intuition, and comes with a burst, then difficulties arise–this thing gives out and [it is] then that "Bugs"–as such little faults and difficulties are called–show themselves and months of intense watching, study and labor are requisite before commercial success or failure is certainly reached.

The quoted paragraph above makes clear that only very exceptionally does any new design fully succeed. The same applies to R scripts as well as any other non-trivial piece of computer code. From this it logically follows that testing and de-bugging are fundamental steps in the development of R scripts and packages. Debugging, as an activity, is outside the scope of this book. However, clear pro-gramming style and good documentation are indispensable for efficient testing and reuse.

Even for scripts used for analyzing a single data set, we need to be confident that the algorithms and their implementation are valid, and able to return correct results. This is true both for scientific reports, expert data-based reports and any data analysis related to assessment of compliance with legislation or regulations. Of course, even in cases when we are not required to demonstrate validity, say for decision making purely internal to a private organization, we will still want to avoid costly mistakes.

The first step in producing reliable computer code is to accept that any code that we write needs to be tested and, if possible, validated. Another important step is to make sure that input is validated within the script and a suitable error produced for bad input (including valid input values falling outside the range that can be reliably handled by the script).

If during testing, or during normal use, a wrong value is returned by a cal-culation, or no value (e.g., the script crashes or triggers a fatal error), debugging consists in finding the cause of the problem. The cause can be either a mistake in the implementation of an algorithm, as well as in the algorithm itself. However, many apparent *bugs* are caused by bad or missing handling of special cases like invalid input values, rounding errors, division by zero, etc., in which a program crashes instead of elegantly issuing a helpful error message.

Diagnosing the source of bugs is, in most cases, like detective work. One uses hunches based on common sense and experience to try to locate the lines of code causing the problem. One follows different *leads* until the case is solved. In most cases, at the very bottom we rely on some sort of divide-and-conquer strategy. For example, we may check the value returned by intermediate calculations until we locate the earliest code statement producing a wrong value. Another common case is when some input values trigger a bug. In such cases it is frequently best to start by testing if different "cases" of input lead to errors/crashes or not. Boundary input values are usually the telltale ones: e.g., for numbers, zero, negative and positive values, very large values, very small values, missing values (NA), vectors of length zero (`numeric()`), etc.

⚠️ **Error messages** When debugging, keep in mind that in some cases a single bug can lead to a whole cascade of error messages. Do also keep in mind that typing mistakes, originating when code is entered through the keyboard, can wreak havock in a script: usually there is little correspondence between the number of error messages and the seriousness of the bug triggering them. When several errors are triggered, start by reading the error message printed first, as later errors can be an indirect consequence of earlier ones.

There are special tools, called debuggers, available, and they help enormously. Debuggers allow one to step through the code, executing one statement at a time, and at each pause, allowing the user to inspect the objects present in the R environment and their values. It is even possible to execute additional statements, say, to modify the value of a variable, while execution is paused. An R debugger is available within RStudio and also through the R console.

When writing your first scripts, you will manage perfectly well, and learn more by running the script one line at a time and when needed temporarily inserting print() statements to "look" at how the value of variables changes at each step. A debugger allows a lot more control, as one can "step in" and "step out" of function definitions, and set and unset break points where execution will stop, which is especially useful when developing R packages.

When reproducing the examples in this chapter, do keep this section in mind. In addition, if you get stuck trying to find the cause of a bug, do extend your search both to the most trivial of possible causes, and to the least likely ones (such as a bug in a package installed from CRAN or R itself). Of course, when suspecting a bug in code you have not written, it is wise to very carefully read the documentation, as the "bug" may be just in your understanding of what a certain piece of code is expected to do. Also keep in mind that as discussed on page 12, you will be able to find online already-answered questions to many of your likely problems and doubts. For example, Googling for the text of an error message is usually well rewarded.

⚠️ When installing packages from other sources than CRAN (e.g., development versions from GitHub, Bitbucket or R-Forge, or in-house packages) there is no warranty that conflicts will not happen. Packages (and their versions) released through CRAN are regularly checked for inter-compatibility, while packages released through other channels are usually checked against only a few packages.

Conflicts among packages can easily arise, for example, when they use the same names for objects or functions. In addition, many packages use functions defined in packages in the R distribution itself or other independently developed packages by importing them. Updates to depended-upon packages can "break" (make non-functional) the dependent packages or parts of them. The rigorous testing by CRAN detects such problems in most cases when package revisions are submitted, forcing package maintainers to fix problems before distribution through CRAN is possible. However, if you use other repositories, I recommend that you make sure that revised (especially if under develop-ment) versions do work with your own script, before their use in "production" (important) data analyses.

3.3   Control of execution flow

We give the name *control of execution statements* to those statements that allow the execution of sections of code when a certain dynamically computed condition is TRUE. Some of the control of execution flow statements, function like *ON-OFF switches* for program statements. Others allow statements to be executed repeat-edly while or until a condition is met, or until all members of a list or a vector are processed.

These *control of execution statements* can be also used at the R console, but it is usually awkward to do so as they can extend over several lines of text. In simple scripts, the *flow of execution* can be fixed and linear from the first to the last statement in the script. *Control of execution statements* allow flexibility, as they allow conditional execution and/or repeated execution of statements. The part of the script conditionally executed can be a simple or a compound code statement providing a lot of flexibility. As we will see next, a compound statement can include multiple simple or nested compound statements.

3.3.1   Compound statements

First of all, we need to consider compound statements. Individual statements can be grouped into compound statements by enclosed them in curly braces.

```
print("A")
## [1] "A"

{
  print("B")
  print("C")
}
## [1] "B"
## [1] "C"
```

The grouping of the last two statements above is of no consequence by itself, but grouping becomes useful when used together with control-of-execution con-structs.

3.3.2   Conditional execution

Conditional execution allows handling different values, such as negative and non-negative values, differently within a script. This is achieved by evaluating or not (i.e., switching ON and OFF) parts of a script based on the result returned by a logical expression. This expression can also be a *flag*—i.e., a `logical` variable set manually, preferable near the top of the script. Use of flags is most useful when switching between two script behaviors depends on multiple sections of code. A frequent use case for flags is jointly enabling and disabling printing of output from multiple statements scattered in over a long script.

R has two types of *if* statements, non-vectorized and vectorized. We will start with the non-vectorized one, which is similar to what is available in most other computer programming languages. We start with toy examples demonstrating how *if* and *if-else* statements work. Later we will see examples closer to real use cases.

3.3.2.1 Non-vectorized `if`, `else` and `switch`

The `if` construct "decides," depending on a `logical` value, whether the next code statement is executed (if TRUE) or skipped (if FALSE).

```
flag <- TRUE
if (flag) print("Hello!")
## [1] "Hello!"
```

> Play with the code above by changing the value assigned to variable `flag`, FALSE, NA, and `logical(0)`.
>
> In the example above we use variable `flag` as the *condition*.
>
> Nothing in the R language prevents this condition from being a `logical` constant. Explain why `if (TRUE)` in the syntactically-correct statement below is of no practical use.
>
> ```
> if (TRUE) print("Hello!")
> ## [1] "Hello!"
> ```

Conditional execution is much more useful than what could be expected from the previous example, because the statement whose execution is being controlled can be a compound statement of almost any length or complexity. A very simple example follows.

```
printing <- TRUE
if (printing) {
  print("A")
  print("B")
}
## [1] "A"
## [1] "B"
```

The condition passed as an argument to `if`, enclosed in parentheses, can be anything yielding a `logical` vector, however, as this condition is *not* vectorized, only the first element will be used and a warning issued if longer than one.

```
a <- 10.0
if (a < 0.0) print("'a' is negative") else print("'a' is not negative")
## [1] "'a' is not negative"

print("This is always printed")
## [1] "This is always printed"
```

As can be seen above, the statement immediately following if is executed if the condition returns TRUE and that following else is executed if the condition returns FALSE. Statements after the conditionally executed if and else statements are always executed, independently of the value returned by the condition.

Play with the code in the chunk above by assigning different numeric vectors to a.

Do you still remember the rules about continuation lines?

```
# 1
a <- 1
if (a < 0.0) print("'a' is negative") else print("'a' is not negative")
## [1] "'a' is not negative"
```

Why does the statement below (not evaluated here) trigger an error while the one above does not?

```
# 2 (not evaluated here)
if (a < 0.0) print("'a' is negative")
else print("'a' is not negative")
```

How do the continuation line rules apply when we add curly braces as shown below.

```
# 1
a <- 1
if (a < 0.0) {
    print("'a' is negative")
  } else {
    print("'a' is not negative")
  }
## [1] "'a' is not negative"
```

In the example above, we enclosed a single statement between each pair of curly braces, but as these braces create compound statements, multiple statements could have been enclosed between each pair.

Play with the use of conditional execution, with both simple and compound statements, and also think how to combine `if` and `else` to select among more than two options.

In R, the value returned by any compound statement is the value returned by the last simple statement executed within the compound one. This means that we can assign the value returned by an `if` and `else` statement to a variable. This style is less frequently used, but occasionally can result in easier-to-understand scripts.

```
a <- 1
my.message <-
  if (a < 0.0) "'a' is negative" else "'a' is not negative"
print(my.message)
## [1] "'a' is not negative"
```

Study the conversion rules between `numeric` and `logical` values, run each of the statements below, and explain the output based on how type conversions are interpreted, remembering the difference between *floating-point numbers* as implemented in computers and *real numbers* ($\mathbb{R}$) as defined in mathematics.

```
if (0) print("hello")
if (-1) print("hello")
if (0.01) print("hello")
if (1e-300) print("hello")
if (1e-323) print("hello")
if (1e-324) print("hello")
if (1e-500) print("hello")
if (as.logical("true")) print("hello")
if (as.logical(as.numeric("1"))) print("hello")
if (as.logical("1")) print("hello")
if ("1") print("hello")
```

Hint: if you need to refresh your understanding of the type conversion rules, see section 2.9 on page 42.

In addition to `if()`, there is in R a `switch()` statement, which we describe next. It can be used to select among *cases*, or several alternative statements, based on an expression evaluating to a `numeric` or a `character` value of length equal to one. The switch statement returns a value, the value returned by the statement corresponding to the matching switch value, or the default if there is no match and a default return value has been defined in the code.

```
my.object <- "two"
b <- switch(my.object,
            one = 1,
            two = 1 / 2,
            three = 1 / 4,
            0
)
b
## [1] 0.5
```

> 📖 Do play with the use of the switch statement. Look at the documentation for `switch()` using `help(switch)` and study the examples at the end of the help page.

The `switch()` statement can substitute for chained `if else` statements when all the conditions are comparisons against different constant values, resulting in more concise and clear code.

3.3.2.2 Vectorized `ifelse()`

Vectorized *ifelse* is a peculiarity of the R language, but very useful for writing concise code that may execute faster than logically equivalent but not vectorized code. Vectorized conditional execution is coded by means of *function* `ifelse()` (written as a single word). This function takes three arguments: a `logical` vector usually the result of a test (parameter `test`), a result vector for TRUE cases (parameter yes), and a result vector for FALSE cases (parameter no). At each index position along the vectors, the value included in the returned vector is taken from yes if `test` is TRUE and from no if `test` is FALSE. All three arguments can be any R statement returning the required vector. In the case of vectors passed as arguments to parameters yes and no, recycling will take place if they are shorter than the logical vector passed as argument to `test`. No recycling ever applies to `test`, even if yes and/or no are longer than `test`. It is customary to pass arguments to `ifelse` by position. We give a first example with named arguments to clarify the use of the function.

```
my.test <- c(TRUE, FALSE, TRUE, TRUE)
ifelse(test = my.test, yes = 1, no = -1)
## [1]  1 -1  1  1
```

In practice, the most common idiom is to have as an argument passed to `test`, the result of a comparison calculated on the fly. In the first example we compute the absolute values for a vector, equivalent to that returned by R function `abs()`.

```
nums <- -3:+3
ifelse(nums < 0, -nums, nums)
## [1] 3 2 1 0 1 2 3
```

> 📖 Some additional examples to play with, with a few surprises. Study the examples below until you understand why returned values are what they are. In addition, create your own examples to test other possible cases. In other words, play with the code until you fully understand how `ifelse` works.
>
> ```
> a <- 1:10
> ifelse(a > 5, 1, -1)
> ifelse(a > 5, a + 1, a - 1)
> ifelse(any(a > 5), a + 1, a - 1) # tricky
> ifelse(logical(0), a + 1, a - 1) # even more tricky
> ifelse(NA, a + 1, a - 1) # as expected
> ```

Hint: if you need to refresh your understanding of `logical` values and Boolean algebra see section 2.4 on page 29.

⚠️ In the case of `ifelse()`, the length of the returned value is determined by the length of the logical vector passed as an argument to its first formal parameter (named `test`)! A frequent mistake is to use a condition that returns a `logical` vector of length one, expecting that it will be recycled because arguments passed to the other formal parameters (named `yes` and `no`) are longer. However, no recycling will take place, resulting in a returned value of length one, with the remaining elements of the vectors passed to `yes` and `no` being discarded. Do try this by yourself, using logical vectors of different lengths. You can start with the examples below, making sure you understand why the returned values are what they are.

```
ifelse(TRUE, 1:5, -5:-1)
## [1] 1
```

```
ifelse(FALSE, 1:5, -5:-1)
## [1] -5
```

```
ifelse(c(TRUE, FALSE), 1:5, -5:-1)
## [1]  1 -4
```

```
ifelse(c(FALSE, TRUE), 1:5, -5:-1)
## [1] -5  2
```

```
ifelse(c(FALSE, TRUE), 1:5, 0)
## [1] 0 2
```

📖 Write, using `ifelse()`, a single statement to combine numbers from the two vectors a and b into a result vector d, based on whether the corresponding value in vector c is the character "a" or "b". Then print vector d to make the result visible.

```
a <- -10:-1
b <- +1:10
c <- c(rep("a", 5), rep("b", 5))
# your code
```

If you do not understand how the three vectors are built, or you cannot guess the values they contain by reading the code, print them, and play with the arguments, until you understnd what each parameter does. Also use `help(rep)` and/or `help(ifelse)` to access the documentation.

3.3.3   Iteration

We give the name *iteration* to the process of repetitive execution of a program statement (simple or compound)—e.g., *computed by iteration*. We use the same word,
iteration, to name each one of these repetitions of the execution of a statement—
e.g., the second iteration.

The section of computer code being executed multiple times, forms a loop (a
closed path). Most loops contain a condition that determines when the flow of execution will exit the loop and continue at the next statement following the loop.
In R three types of iteration loops are available: those using `for`, `while` and `repeat`
constructs. They differ in how much flexibility they provide with respect to the values they iterate over, and how the condition that terminates the iteration is tested.
When the same algorithm can be implemented with more than one of these constructs, using the least flexible of them usually results in the easiest to understand
R scripts. In R, rather frequently, explicit loops as described in this section can be
replaced advantageously by calls to the *apply* functions described in section 3.4
on page 108.

3.3.3.1   for loops

The most frequently used type of loop is a `for` loop. These loops work in R on lists
or vectors of values to act upon.

```
b <- 0
for (a in 1:5) b <- b + a
b
## [1] 15

b <- sum(1:5) # built-in function (faster)
b
## [1] 15
```

Here the statement b <- b + a is executed five times, with variable a sequentially taking each of the values in 1:5. Instead of a simple statement used here, a
compound statement could also have been used for the body of the `for` loop.

⚠️ It is important to note that a list or vector of length zero is a valid argument to `for()`, that triggers no error, but skips the statements in the loop
body.

Some examples of use of `for` loops—and of how to avoid their use.

```
a <- c(1, 4, 3, 6, 8)
for(x in a) {print(x*2)} # print is needed!
## [1] 2
## [1] 8
## [1] 6
## [1] 12
## [1] 16
```

A call to `for` does not return a value. We need to assign values to an object so that they are not lost. If we print at each iteration the value of this object, we can follow how the stored value changes. Printing allows us to see, how the vector grows in length, unless we create a long-enough vector before the start of the loop.

```
b <- for(x in a) {x*2}
b
## NULL

b <- numeric()
for(i in seq(along.with = a)) {
  b[i] <- a[i]^2
  print(b)
}
## [1] 1
## [1]  1 16
## [1]  1 16  9
## [1]  1 16  9 36
## [1]  1 16  9 36 64

b
## [1]  1 16  9 36 64

# runs faster if we first allocate a long enough vector
b <- numeric(length(a))
for(i in seq(along.with = a)) {
  b[i] <- a[i]^2
  print(b)
}
## [1] 1 0 0 0 0
## [1]  1 16  0  0  0
## [1]  1 16  9  0  0
## [1]  1 16  9 36  0
## [1]  1 16  9 36 64

b
## [1]  1 16  9 36 64

# a vectorized expression is simplest and fastest
b <- a^2
b
## [1]  1 16  9 36 64
```

In the previous chunk we used `seq(along.with = a)` to build a new numeric vector with a sequence of the same length as vector a. Using this *idiom* is best as it ensures that even the case when a is an *empty* vector of length zero will be handled correctly, with `numeric(0)` assigned to b.

Look at the results from the above examples, and try to understand where the returned value comes from in each case. In the code chunk above, `print()` is used within the *loop* to make intermediate values visible. You can add additional `print()` statements to visualize other variables, such as i, or run parts of the code, such as `seq(along.with = a)`, by themselves.

In this case, the code examples trigger no errors or warnings, but the same

approach can be used for debugging syntactically correct code that does not
return the expected results.

In the examples above we show the use of seq() passing a vector as
an argument to its parameter along.with. Run the examples below and explain
why the two approaches are equivalent only when the length of a is one or
more. Find the answer by assigning to a, vectors of different lengths, including
zero (using a <- numeric(0)).

```r
b <- numeric(length(a))
for(i in seq(along.with = a)) {
  b[i] <- a[i]^2
}
print(b)

c <- numeric(length(a))
for(i in 1:length(a)) {
  c[i] <- a[i]^2
}
print(c)
```

for loops as described above, in the absence of errors, have statically pre-
dictable behavior. The compound statement in the loop will be executed once
for each member of the vector or list. Special cases may require the alteration
of the normal flow of execution in the loop. Two cases are easy to deal with,
one is stopping iteration early, which we can do with break(), and another is
jumping ahead to the start of the next iteration, which we can do with next().

3.3.3.2 while loops

while loops are frequently useful, even if not as frequently used as for loops.
Instead of a list or vector, they take a logical argument, which is usually an expression, but which can also be a variable.

```r
a <- 2
while (a < 50) {
  print(a)
  a <- a^2
}
## [1] 2
## [1] 4
## [1] 16

print(a)
## [1] 256
```

Make sure that you understand why the final value of a is larger than 50.

The statements above can be simplified to:

```
a <- 2
print(a)
while (a < 50) {
  print(a <- a^2)
}
```

Explain why this works, and how it relates to the support in R of *chained* assignments to several variables within a single statement like the one below.

```
a <- b <- c <- 1:5
a
```

Explain why a second print(a) has been added before while(). Hint: experiment if necessary.

while loops as described above will terminate when the condition tested is FALSE. In those cases that require stopping iteration based on an additional test condition within the compound statement, we can call break() in the body of an if or else statement.

3.3.3.3 repeat loops

The repeat construct is less frequently used, but adds flexibility as termination will always depend on a call to break(), which can be located anywhere within the compound statement that forms the body of the loop. To achieve conditional end of iteration, function break() must be called, as otherwise, iteration in a repeat loop will not stop.

```
a <- 2
repeat{
  print(a)
  if (a > 50) break()
  a <- a^2
}
## [1] 2
## [1] 4
## [1] 16
## [1] 256
```

> 📇 Please explain why the example above returns the values it does. Use the approach of adding `print()` statements, as described on page 101.

> ☕ Although `repeat` loop constructs are easier to read if they have a single condition resulting in termination of iteration, it is allowed by the R language for the compound statement in the body of a loop to contain more than one call to `break()`, each within a different `if` or `else` statement.

3.3.4 Explicit loops can be slow in R

If you have written programs in other languages, it will feel natural to you to use loops (`for`, `while`, `repeat`) for many of the things for which in R one would normally use vectorization. In R, using vectorization whenever possible keeps scripts shorter and easier to understand (at least for those with experience in R). More importantly, as R is an interpreted language, vectorized arithmetic tends to be much faster than the use of explicit iteration. In recent versions of R, byte-compilation is used by default and loops may be compiled on the fly, which relieves part of the burden of repeated interpretation. However, even byte-compiled loops are usually slower to execute than efficiently coded vectorized functions and operators.

Execution speed needs to be balanced against the effort invested in writing faster code. However, using vectorization and specific R functions requires little effort once we are familiar with them. The simplest way of measuring the execution time of an R expression is to use function `system.time()`. However, the returned time is in seconds and consequently the expression must take long enough to execute for the returned time to have useful resolution. See package 'microbenchmark' for tools for benchmarking code with better time resolution.

```
system.time({a <- numeric()
            for (i in 1:1000000) {
              a[i] <- i / 1000
              }
            })
##    user  system elapsed
##    0.44    0.03    0.48
```

> ☕ Whenever working with large data sets, or many similar data sets, we will need to take performance into account. As vectorization usually also makes code simpler, it is good style to use vectorization whenever possible. For operations that are frequently used, R includes specific functions. It is thus important to consider not only vectorization of arithmetic but also check for the availability of performance-optimized functions for specific cases. The results from running the code examples in this box are not included, because they are

the same for all chunks. Here we are interested in the execution time, and we leave this as an exercise.

```
a <- rnorm(10^7) # 10 000 0000 pseudo-random numbers


# b <- numeric()
b <- numeric(length(a)-1) # pre-allocate memory
i <- 1
while (i < length(a)) {
  b[i] <- a[i+1] - a[i]
  print(b)
  i <- i + 1
}
b


# b <- numeric()
b <- numeric(length(a)-1) # pre-allocate memory
for(i in seq(along.with = b)) {
  b[i] <- a[i+1] - a[i]
  print(b)
}
b


# although in this case there were alternatives, there
# are other cases when we need to use indexes explicitly
b <- a[2:length(a)] - a[1:length(a)-1]
b


# or even better
b <- diff(a)
b
```

Execution time can be obtained with `system.time()`. For a vector of ten million numbers, the `for` loop above takes 1.1 s and the equivalent `while` loop 2.0 s, the vectorized statement using indexing takes 0.2 s and function `diff()` takes 0.1 s. The `for` loop without pre-allocation of memory to b takes 3.6 s, and the equivalent while loop 4.7 s—i.e., the fastest execution time was more than 40 times faster than the slowest one. (Times for R 3.5.1 on my laptop under Windows 10 x64.)

3.3.5 Nesting of loops

All the execution-flow control statements seen above can be nested. We will show an example with two `for` loops. We first create a matrix of data to work with:

```
A <- matrix(1:50, 10)
A
##       [,1] [,2] [,3] [,4] [,5]
```

```
## [1,]    1  11  21  31  41
## [2,]    2  12  22  32  42
## [3,]    3  13  23  33  43
## [4,]    4  14  24  34  44
## [5,]    5  15  25  35  45
## [6,]    6  16  26  36  46
## [7,]    7  17  27  37  47
## [8,]    8  18  28  38  48
## [9,]    9  19  29  39  49
## [10,]  10  20  30  40  50
```

```
row.sum <- numeric()
for (i in 1:nrow(A)) {
  row.sum[i] <- 0
  for (j in 1:ncol(A))
    row.sum[i] <- row.sum[i] + A[i, j]
}
print(row.sum)
## [1] 105 110 115 120 125 130 135 140 145 150
```

The code above is very general, it will work with any two-dimensional matrix with at least one column and one row. However, sometimes we need more specific calculations. A[1, 2] selects one cell in the matrix, the one on the first row of the second column. A[1,] selects row one, and A[, 2] selects column two. In the example above, the value of i changes for each iteration of the outer loop. The value of j changes for each iteration of the inner loop, and the inner loop is run in full for each iteration of the outer loop. The inner loop index j changes fastest.

---

1) Modify the code in the example in the last chunk above so that it sums the values only in the first three columns of A, 2) modify the same example so that it sums the values only in the last three rows of A, 3) modify the code so that matrices with dimensions equal to zero (as reported by ncol() and nrow()).

Will the code you wrote continue working as expected if the number of rows in A changed? What if the number of columns in A changed, and the required results still needed to be calculated for relative positions? What would happen if A had fewer than three columns? Try to think first what to expect based on the code you wrote. Then create matrices of different sizes and test your code. After that, think how to improve the code, so that wrong results are not produced.

If the total number of iterations is large and the code executed at each iteration runs fast, the overhead added by the loop code can make a big contribution to the total running time of a script. When dealing with nested loops, as the inner loop is executed most frequently, this is the best place to look for ways of reducing execution time. In this example, vectorization can be achieved easily for the inner loop, as R has a function `sum()` which returns the sum of a vector passed as its argument. Replacing the inner loop by an efficient function can be expected to improve performance significantly.

```
row.sum <- numeric(nrow(A)) # faster
for (i in 1:nrow(A)) {
  row.sum[i] <- sum(A[i, ])
}
print(row.sum)
##  [1] 105 110 115 120 125 130 135 140 145 150
```

`A[i, ]` selects row `i` and all columns. Reminder: in R the row index comes first.

Both explicit loops can be eliminated if we use an *apply* function, such as `apply()`, `lapply()` or `sapply()`, in place of the outer `for` loop. See section 3.4 below for details on the use of the different *apply* functions.

```
row.sum <- apply(A, MARGIN = 1, sum) # MARGIN=1 indicates rows
print(row.sum)
##  [1] 105 110 115 120 125 130 135 140 145 150
```

Calculating row sums is a frequent operation, so R has a built-in function for this. As earlier with `diff()`, it is always worthwhile to check if there is an existing R function, optimized for performance, capable of doing the computations we need. In this case, using `rowSums()` simplifies the nested loops into a single function call, both improving performance and readability.

```
rowSums(A)
##  [1] 105 110 115 120 125 130 135 140 145 150
```

1) How would you change this last example, so that only the last three columns are added up? (Think about use of subscripts to select a part of the matrix.) 2) To obtain column sums, one could modify the nested loops (think how), transpose the matrix and use `rowSums()` (think how), or look up if there is in R a function for this operation. A good place to start is with `help(rowSums)` as similar functions may share the same help page, or at least be listed in the "See also" section. Do try this, and explore other help pages in search for some function you may find useful in the analysis of your own data.

3.3.5.1 Clean-up

Sometimes we need to make sure that clean-up code is executed even if the execution of a script or function is aborted by the user or as a result of an error condition. A typical example is a script that temporarily sets a disk folder as the working directory or uses a file as temporary storage. Function `on.exit()` can be used to record that a user supplied expression needs to be executed when the current function, or a script, exits. Function `on.exit()` can also make code easier to read as it keeps creation and clean-up next to each other in the body of a function or in the listing of a script.

```
file.create("temp.file")
## [1] TRUE

on.exit(file.remove("temp.file"))
# code that makes use of the file goes here
```

3.4 *Apply* functions

Apply functions apply a function passed as an argument to parameter FUN or equivalent, to elements in a collection of R objects passed as an argument to parameter x or equivalent. Collections to which FUN is to be applied can be vectors, lists, data frames, matrices or arrays. As long as the operations to be applied are *independent—i.e., the results from one iteration are not used in another iteration—* apply functions can replace `for`, `while` or `repeat` loops.

The different *apply* functions in base R differ in the class of the values they accept for their x parameter, the class of the object they return and/or the class of the value returned by the applied function. `lapply()` and `sapply()` expect a `vector` or `list` as an argument passed through x. `lapply()` returns a `list` or an `array`; and `vapply()` always *simplifies* its returned value into a vector, while `sapply()` does the simplification according to the argument passed to its `simplify` parameter. All these *apply* functions can be used to apply an R function that returns a value of the same or a different class as its argument. In the case of `apply()` and `lapply()` not even the length of the values returned for each member of the collection passed as an argument, needs to be consistent. In summary, `apply()` is used to apply a function to the elements along a dimension of an object that has two or more *dimensions*, and `lapply()` and `sapply()` are used to apply a function to the members of a vector or list. `apply()` returns an array or a list or a vector depending on the size, and consistency in length and class among the values returned by the applied function.

3.4.1   Applying functions to vectors and lists

We first exemplify the use of `lapply()`, `sapply()` and `vapply()`. In the chunks below we apply a user-defined function to a vector.

⚠ A constraint on the function to be applied is that the member object will be always passed as an argument to the first parameter of the applied function.

```
set.seed(123456) # so that a.vector does not change
a.vector <- runif(6) # A short vector as input to keep output short
str(a.vector)
##  num [1:6] 0.798 0.754 0.391 0.342 0.361 ...
```

```
my.fun <- function(x, k) {log(x) + k}
```

```
z <- lapply(x = a.vector, FUN = my.fun, k = 5)
str(z)
## List of 6
##  $ : num 4.77
##  $ : num 4.72
##  $ : num 4.06
##  $ : num 3.93
##  $ : num 3.98
##  $ : num 3.38
```

```
z <- sapply(x = a.vector, FUN = my.fun, k = 5)
str(z)
##  num [1:6] 4.77 4.72 4.06 3.93 3.98 ...
```

```
z <- sapply(x = a.vector, FUN = my.fun, k = 5, simplify = FALSE)
str(z)
## List of 6
##  $ : num 4.77
##  $ : num 4.72
##  $ : num 4.06
##  $ : num 3.93
##  $ : num 3.98
##  $ : num 3.38
```

We can see above that the computed results are the same in the three cases, but the class and structure of the objects returned differ.

Anonymous functions can be defined on the fly and passed to FUN, allowing us to re-write the examples above more concisely (only the second one shown).

```
z <- sapply(x = a.vector, FUN = function(x, k) {log(x) + k}, k = 5)
str(z)
##  num [1:6] 4.77 4.72 4.06 3.93 3.98 ...
```

Of course, as discussed in section 3.3.4 on page 104, when suitable vectorized functions are available, their use should be preferred. On the other hand, even if *apply* functions are usually not as fast as vectorized functions, they are faster than the equivalent for() loops.

```
z <- log(a.vector) + 5
str(z)
##  num [1:6] 4.77 4.72 4.06 3.93 3.98 ...
```

Function `vapply()` can be safer to use as the mode of returned values is enforced. Here is a possible way of obtaining means and variances across member vectors at each vector index position from a list of vectors. These could be called *parallel* means and variances. The argument passed to FUN.VALUE provides a template for the type of the return value and its organization into rows and columns. Notice that the rows in the output are now named according to the names in FUN.VALUE.

We first use `lapply()` to create the object a.list containing artificial data. One or more additional *named* arguments can be passed to the function to be applied.

```
set.seed(123456)
a.list <- lapply(rep(4, 5), rnorm, mean = 10, sd = 1)
str(a.list)
## List of 5
##  $ : num [1:4] 10.83 9.72 9.64 10.09
##  $ : num [1:4] 12.3 10.8 11.3 12.5
##  $ : num [1:4] 11.17 9.57 9 8.89
##  $ : num [1:4] 9.94 11.17 11.05 10.06
##  $ : num [1:4] 9.26 10.93 11.67 10.56
```

We define the function that we will apply, a function that returns a numeric vector of length 2.

```
mean_and_sd <- function(x, na.rm = FALSE) {
     c(mean(x, na.rm = na.rm),  sd(x, na.rm = na.rm))
  }
```

We next use `vapply()` to apply our function to each member vector of the list.

```
values <- vapply(x = a.list,
                 FUN = mean_and_sd,
                 FUN.VALUE = c(mean = 0,  sd = 0),
                 na.rm = TRUE)
class(values)
## [1] "matrix" "array"

values
##            [,1]       [,2]      [,3]       [,4]      [,5]
## mean 10.0725427 11.7254442 9.657997 10.5573814 10.605846
## sd    0.5428149  0.7844356 1.050663  0.6460881  1.005676
```

As explained in section 2.14 on page 66, class `data.frame` is derived from class `list`. Apply function `mean_and_sd()` defined above to the data frame cars included as example data in R. The aim is to obtain the mean and standard deviation for each column.

3.4.2   Applying functions to matrices and arrays

In the next example we use `apply()` and `mean()` to compute the mean for each column of matrix `a.matrix`. In R the dimensions of a matrix, rows and columns, over which a function is applied are called *margins*. The argument passed to parameter MARGIN determines over which margin the function will be applied. If the function is applied to individual rows, we say that we operate on the first margin, and if the function is applied to individual columns, over the second margin. Arrays can have many dimensions, and consequently more margins. In the case of arrays with more than two dimensions, it is possible and useful to apply functions over multiple margins at once.

> ⚠ A constraint on the function to be applied is that the vector or "slice" will always be passed as a positional argument to the first formal parameter of the applied function.

```
a.matrix <- matrix(runif(100), ncol = 10)
z <- apply(a.matrix, MARGIN = 1, FUN = mean)
str(z)
##  num [1:10] 0.247 0.404 0.537 0.5 0.504 ...
```

> 📶 Modify the example above so that it computes row means instead of column means.

> 📶 Look up the help pages for `apply()` and `mean()` and study them until you understand how additional arguments can be passed to the applied function. Can you guess why `apply()` was designed to have parameter names fully in uppercase, something very unusual for R code style?

If we apply a function that returns a value of the same length as its input, then the dimensions of the value returned by `apply()` are the same as those of its input. We use, in the next examples, a "no-op" function that returns its argument unchanged, so that input and output can be easily compared.

```
a.small.matrix <- matrix(rnorm(6, mean = 10, sd = 1), ncol = 2)
a.small.matrix <- round(a.small.matrix, digits = 1)
a.small.matrix
##       [,1] [,2]
## [1,] 11.3 10.4
## [2,] 10.6  8.6
## [3,]  8.2 11.0
```

```
no_op.fun <- function(x) {x}
```

```
z <- apply(x = a.small.matrix, MARGIN = 2, FUN = no_op.fun)
class(z)
## [1] "matrix" "array"
```

```
z
##       [,1] [,2]
## [1,] 11.3 10.4
## [2,] 10.6  8.6
## [3,]  8.2 11.0
```

In the chunk above, we passed MARGIN = 2, but if we pass MARGIN = 1, we get a return value that is transposed! To restore the original layout of the matrix we can transpose the result with function t().

```
z <- apply(x = a.small.matrix, MARGIN = 1, FUN = no_op.fun)
z
##       [,1] [,2] [,3]
## [1,] 11.3 10.6  8.2
## [2,] 10.4  8.6 11.0
```

```
t(z)
##       [,1] [,2]
## [1,] 11.3 10.4
## [2,] 10.6  8.6
## [3,]  8.2 11.0
```

A more realistic example, but difficult to grasp without seeing the toy examples shown above, is when we apply a function that returns a value of a different length than its input, but longer than one. When we compute column summaries (MARGIN = 2), a matrix is returned, with each column containing the summaries for the corresponding column in the original matrix (a.small.matrix). In contrast, when we compute row summaries (MARGIN = 1), each column in the returned matrix contains the summaries for one row in the original array. What happens is that by using apply() the dimension of the original matrix or array over which we compute summaries "disappears." Consequently, given how matrices are stored in R, when columns collapse into a single value, the rows become columns. After this, the vectors returned by the applied function, are stored as rows.

```
mean_and_sd <- function(x, na.rm = FALSE) {
    c(mean(x, na.rm = na.rm),  sd(x, na.rm = na.rm))
  }
```

```
z <- apply(x = a.small.matrix, MARGIN = 2, FUN = mean_and_sd, na.rm = TRUE)
z
##             [,1]    [,2]
## [1,] 10.033333 10.000
## [2,]  1.625833  1.249
```

```
z <- apply(X = a.small.matrix, MARGIN = 1, FUN = mean_and_sd, na.rm = TRUE)
z
##              [,1]     [,2]     [,3]
## [1,] 10.8500000 9.600000 9.600000
## [2,]  0.6363961 1.414214 1.979899
```

In all examples above, we have used ordinary functions. Operators in R are functions with two formal parameters which can be called using infix notation in expressions—i.e., a + b. By back-quoting their names they can be called using the same syntax as for ordinary functions, and consequently also passed to the FUN parameter of apply functions. A toy example, equivalent to the vectorized operation a.vector + 5 follows. We enclosed operator + in back ticks (`) and pass by name a constant to its second formal parameter (e2 = 5).

```
set.seed(123456) # so that a.vector does not change
a.vector <- runif(10)
z <- sapply(X = a.vector, FUN = `+`, e2 = 5)
str(z)
##  num [1:10] 5.8 5.75 5.39 5.34 5.36 ...
```

> 📖 **Apply functions vs. loop constructs** Apply functions cannot always replace explicit loops as they are less flexible. A simple example is the accumulation pattern, where we "walk" through a collection that stores a partial result between iterations. A similar case is a pattern where calculations are done over a "window" that moves at each iteration. The simplest and probably most frequent calculation of this kind is the calculation of differences between successive members. Other examples are moving window summaries such as a moving median (see page 104 for other alternatives to the use of explicit iteration loops).

3.5   Object names and character strings

In all assignment examples before this section, we have used object names included as literal character strings in the code expressions. In other words, the names are "decided" as part of the code, rather than at run time. In scripts or packages, the object name to be assigned may need to be decided at run time and, consequently, be available only as a character string stored in a variable. In this case, function assign() must be used instead of the operators <- or ->. The statements below demonstrate its use.

First using a character constant.

```
assign("a", 9.99)
a
## [1] 9.99
```

Next using a `character` value stored in a variable.

```
name.of.var <- "b"
assign(name.of.var, 9.99)
b
## [1] 9.99
```

The two toy examples above do not demonstrate why one may want to use `assign()`. Common situations where we may want to use character strings to store (future or existing) object names are 1) when we allow users to provide names for objects either interactively or as `character` data, 2) when in a loop we transverse a vector or list of object names, or 3) we construct at runtime object names from multiple character strings based on data or settings. A common case is when we import data from a text file and we want to name the object according to the name of the file on disk, or a character string read from the header at the top of the file.

Another case is when `character` values are the result of a computation.

```
for (i in 1:5) {
    assign(paste("zz_", i, sep = ""), i^2)
}
ls(pattern = "zz_*")
## [1] "zz_1" "zz_2" "zz_3" "zz_4" "zz_5"
```

The complementary operation of *assigning* a name to an object is to *get* an object when we have available its name as a character string. The corresponding function is `get()`.

```
get("a")
## [1] 9.99
```

```
get("b")
## [1] 9.99
```

If we have available a character vector containing object names and we want to create a list containing these objects we can use function `mget()`. In the example below we use function `ls()` to obtain a character vector of object names matching a specific pattern and then collect all these objects into a list.

```
obj_names <- ls(pattern = "zz_*")
obj_lst <- mget(obj_names)
str(obj_lst)
## List of 5
##  $ zz_1: num 1
##  $ zz_2: num 4
##  $ zz_3: num 9
##  $ zz_4: num 16
##  $ zz_5: num 25
```

> Think of possible uses of functions `assign()`, `get()` and `mget()` in scripts you use or could use to analyze your own data (or from other sources). Write a script to implement this, and iteratively test and revise this script until the result produced by the script matches your expectations.

3.6 The multiple faces of loops

To close this chapter, I will mention some advanced aspects of the R language that are useful when writing complex scrips—if you are going through the book sequentially, you will want to return to this section after reading chapters 4 and 5. In the same way as we can assign names to numeric, character and other types of objects, we can assign names to functions and expressions. We can also create lists of functions and/or expressions. The R language has a very consistent grammar, with all lists and vectors behaving in the same way. The implication of this is that we can assign different functions or expressions to a given name, and consequently it is possible to write loops over lists of functions or expressions.

In this first example we use a *character vector of function names*, and use function do.call() as it accepts either character strings or function names as its first argument. We obtain a numeric vector with named members with names matching the function names.

```
x <- rnorm(10)
results <- numeric()
fun.names <- c("mean", "max", "min")
for (f.name in fun.names) {
   results[[f.name]] <- do.call(f.name, list(x))
   }
results
##      mean        max        min
##  0.5453427  2.5026454 -1.1139499
```

When traversing a *list of functions* in a loop, we face the problem that we cannot access the original names of the functions as what is stored in the list are the definitions of the functions. In this case, we can hold the function definitions in the loop variable (f in the chunk below) and call the functions by use of the function call notation (f()). We obtain a numeric vector with anonymous members.

```
results <- numeric()
funs <- list(mean, max, min)
for (f in funs) {
   results <- c(results, f(x))
   }
results
## [1]  0.5453427  2.5026454 -1.1139499
```

We can use a named list of functions to gain full control of the naming of the results. We obtain a numeric vector with named members with names matching the names given to the list members.

```
results <- numeric()
funs <- list(average = mean, maximum = max, minimum = min)
for (f in names(funs)) {
   results[[f]] <- funs[[f]](x)
   }
results
##    average    maximum    minimum
##  0.5453427  2.5026454 -1.1139499
```

Next is an example using model formulas. We use a loop to fit three models, obtaining a list of fitted models. We cannot pass to anova() this list of fitted models, as it expects each fitted model as a separate nameless argument to its ... parameter. We can get around this problem using function do.call() to call anova(). Function do.call() passes the members of the list passed as its second argument as individual arguments to the function being called, using their names if present. anova() expects nameless arguments so we need to remove the names present in results.

```
my.data <- data.frame(x = 1:10, y = 1:10 + rnorm(10, 1, 0.1))
results <- list()
models <- list(linear = y ~ x, linear.orig = y ~ x - 1, quadratic = y ~ x + I(x^2))
for (m in names(models)) {
    results[[m]] <- lm(models[[m]], data = my.data)
    }
str(results, max.level = 1)
## List of 3
##  $ linear      :List of 12
##   ..- attr(*, "class")= chr "lm"
##  $ linear.orig:List of 12
##   ..- attr(*, "class")= chr "lm"
##  $ quadratic  :List of 12
##   ..- attr(*, "class")= chr "lm"

do.call(anova, unname(results))
## Analysis of Variance Table
##
## Model 1: y ~ x
## Model 2: y ~ x - 1
## Model 3: y ~ x + I(x^2)
##   Res.Df      RSS Df Sum of Sq      F     Pr(>F)
## 1      8 0.05525
## 2      9 2.31266 -1   -2.2574 306.19 4.901e-07 ***
## 3      7 0.05161  2    2.2611 153.34 1.660e-06 ***
## ---
## Signif. codes:  0 '***' 0.001 '**' 0.01 '*' 0.05 '.' 0.1 ' ' 1
```

If we had no further use for results we could simply build a list with nameless members by using positional indexing.

```
results <- list()
models <- list(y ~ x, y ~ x - 1, y ~ x + I(x^2))
for (i in seq(along.with = models)) {
    results[[i]] <- lm(models[[i]], data = my.data)
    }
str(results, max.level = 1)
## List of 3
##  $ :List of 12
##   ..- attr(*, "class")= chr "lm"
##  $ :List of 12
##   ..- attr(*, "class")= chr "lm"
##  $ :List of 12
##   ..- attr(*, "class")= chr "lm"

do.call(anova, results)
## Analysis of Variance Table
##
```

```
## Model 1: y ~ x
## Model 2: y ~ x - 1
## Model 3: y ~ x + I(x^2)
##   Res.Df     RSS Df Sum of Sq      F    Pr(>F)
## 1      8 0.05525
## 2      9 2.31266 -1   -2.2574 306.19 4.901e-07 ***
## 3      7 0.05161  2    2.2611 153.34 1.660e-06 ***
## ---
## Signif. codes:  0 '***' 0.001 '**' 0.01 '*' 0.05 '.' 0.1 ' ' 1
```

3.6.1   Further reading

For further readings on the aspects of R discussed in the current chapter, I suggest the books *The Art of R Programming: A Tour of Statistical Software Design* (Matloff) and *Advanced R* (Wickham).

4

The R language: Statistics

The purpose of computing is insight, not numbers.

Richard W. Hamming
Numerical Methods for Scientists and Engineers, 1987

4.1 Aims of this chapter

This chapter aims to give the reader only a quick introduction to statistics in base R, as there are many good texts on the use of R for different kinds of statistical analyses (see further reading on page 161). Although many of base R's functions are specific to given statistical procedures, they use a particular approach to model specification and for returning the computed values that can be considered a part of the R language. Here you will learn the approaches used in R for calculating statistical summaries, generating (pseudo-)random numbers, sampling, fitting models and carrying out tests of significance. We will use linear correlation, *t*-test, linear models, generalized linear models, non-linear models and some simple multivariate methods as examples. My aim is teaching how to specify models, contrasts and data used, and how to access different components of the objects returned by the corresponding fit and summary functions.

4.2 Statistical summaries

Being the main focus of the R language in data analysis and statistics, R provides functions for both simple and complex calculations, going from means and variances to fitting very complex models. Below are examples of functions implementing the calculation of the frequently used data summaries mean or average (`mean()`), variance (`var()`), standard deviation (`sd()`), median (`median()`), mean absolute deviation (`mad()`), mode (`mode()`), maximum (`max()`), minimum (`min()`), range (`range()`), quantiles (`quantile()`), length (`length()`), and all-encompassing summaries (`summary()`). All these methods accept numeric vectors and matrices as an argument. Some of them also have definitions for other classes such as data frames in the case of `summary()`. (The R language does not define a function for calculation of the standard error of the mean. Please, see section 5.3.1 on page 168 for how to define your own.)

```
x <- 1:20
mean(x)
var(x)
sd(x)
median(x)
mad(x)
mode(x)
max(x)
min(x)
range(x)
quantile(x)
length(x)
summary(x)
```

> In contrast to many other examples in this book, the summaries computed with the code in the previous chunk are not shown. You should *run* them, using vector x as defined above, and then play with other real or artificial data that you may find interesting.

By default, if the argument contains NAs these functions return NA. The logic behind this is that if one value exists but is unknown, the true result of the computation is unknown (see page 25 for details on the role of NA in R). However, an additional parameter called `na.omit` allows us to override this default behavior by requesting any NA in the input to be omitted (or discarded) before calculation,

```
x <- c(1:20, NA)
mean(x)
## [1] NA

mean(x, na.omit = TRUE)
## [1] NA
```

4.3   Distributions

Density, distribution functions, quantile functions and generation of pseudo-random values for several different distributions are part of the R language. Entering `help(Distributions)` at the R prompt will open a help page describing all the distributions available in base R. In what follows we use the Normal distribution for the examples, but with slight differences in their parameters the functions for other theoretical distributions follow a consistent naming pattern. For each distribution the different functions contain the same "root" in their names: `norm` for the normal distribution, `unif` for the uniform distribution, and so on. The "head" of the name indicates the type of values returned: "d" for density, "q" for quantile, "r" (pseudo-)random numbers, and "p" for probabilities.

4.3.1 Density from parameters

Theoretical distributions are defined by mathematical functions that accept parameters that control the exact shape and location. In the case of the Normal distribution, these parameters are the *mean* controlling location and (standard deviation) (or its square, the *variance*) controlling the spread around the center of the distribution.

To obtain a single point from the distribution curve we pass a vector of length one as an argument for x.

```
dnorm(x = 1.5, mean = 1, sd = 0.5)
## [1] 0.4839414
```

To obtain multiple values we can pass a longer vector as an argument. As perusing a long vector of numbers is difficult, we plot the result of the computation as a line (`type = "l"`) that shows that the 50 generated data points give the illusion of a continuous curve.

```
my.x <- seq(from = -1, to = 3, length.out = 50)

my.data <- data.frame(x = my.x,
                      y = dnorm(x = my.x, mean = 1, sd = 0.5))
plot(y~x, data = my.data, type = "l")
```

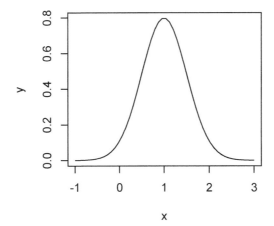

4.3.2   Probabilities from parameters and quantiles

If we have a calculated quantile we can look up the corresponding *p*-value from the Normal distribution. The mean and standard deviation would, in such a case, also be computed from the same observations under the null hypothesis. In the example below, we use invented values for all parameters q, the quantile, mean, and sd, the standard deviation. Use

```r
pnorm(q = 4, mean = 0, sd = 1)
## [1] 0.9999683

pnorm(q = 4, mean = 0, sd = 1, lower.tail = FALSE)
## [1] 3.167124e-05

pnorm(q = 4, mean = 0, sd = 4, lower.tail = FALSE)
## [1] 0.1586553

pnorm(q = c(2, 4), mean = 0, sd = 1, lower.tail = FALSE)
## [1] 2.275013e-02 3.167124e-05

pnorm(q = 4, mean = 0, sd = c(1, 4), lower.tail = FALSE)
## [1] 3.167124e-05 1.586553e-01
```

> ☕ In tests of significance, empirical *z*-values and *t*-values are computed by subtracting from the observed mean for one group or raw quantile, the "expected" mean (possibly a hypothesized theoretical value, the mean of a control condition used as reference, or the mean computed over all treatments under the assumption of no effect of treatments) and then dividing by the standard deviation. Consequently, the *p*-values corresponding to these empirical *z*-values and *t*-values need to be looked up using mean = 0 and sd = 1 when calling pnorm() or pt() respectively. These frequently used values are the defaults.

4.3.3   Quantiles from parameters and probabilities

The reverse computation from that in the previous section is to obtain the quantile corresponding to a known *p*-value. These quantiles are equivalent to the values in the tables used earlier to assess significance.

```r
qnorm(p = 0.01, mean = 0, sd = 1)
## [1] -2.326348

qnorm(p = 0.05, mean = 0, sd = 1)
## [1] -1.644854

qnorm(p = 0.05, mean = 0, sd = 1, lower.tail = FALSE)
## [1] 1.644854
```

⚠ Quantile functions like `qnorm()` and probability functions like `pnorm()` always do computations based on a single tail of the distribution, even though it is possible to specify which tail we are interested in. If we are interested in obtaining simultaneous quantiles for both tails, we need to do this manually. If we are aiming at quantiles for $P = 0.05$, we need to find the quantile for each tail based on $P/2 = 0.025$.

```
qnorm(p = 0.025, mean = 0, sd = 1)
## [1] -1.959964
```

```
qnorm(p = 0.025, mean = 0, sd = 1, lower.tail = FALSE)
## [1] 1.959964
```

We see above that in the case of a symmetric distribution like the Normal, the quantiles in the two tails differ only in sign. This is not the case for asymmetric distributions.

When calculating a p-value from a quantile in a test of significance, we need to first decide whether a two-sided or single-sided test is relevant, and in the case of a single sided test, which tail is of interest. For a two-sided test we need to multiply the returned value by 2.

```
pnorm(q = 4, mean = 0, sd = 1) * 2
## [1] 1.999937
```

4.3.4 "Random" draws from a distribution

True random sequences can only be generated by physical processes. All so-called "random" sequences of numbers generated by computation are really deterministic although they share some properties with true random sequences (e.g., in relation to autocorrelation). It is possible to compute not only pseudo-random draws from a uniform distribution but also from the Normal, t, F and other distributions. Parameter `n` indicates the number of values to be drawn, or its equivalent, the length of the vector returned.

```
rnorm(5)
## [1] -0.8248801  0.1201213 -0.4787266 -0.7134216  1.1264443
```

```
rnorm(n = 10, mean = 10, sd = 2)
##  [1] 12.394190  9.697729  9.212345 11.624844 12.194317 10.257707 10.082981
##  [8] 10.268540 10.792963  7.772915
```

📖 Edit the examples in sections 4.3.2, 4.3.3 and 4.3.4 to do computations based on different distributions, such as Student's t, F or uniform.

📖 It is impossible to generate truly random sequences of numbers by means of a deterministic process such as a mathematical computation. "Random numbers" as generated by R and other computer programs are *pseudo random numbers*, long deterministic series of numbers that resemble random draws. Random number generation uses a *seed* value that determines where in the series we start. The usual way of automatically setting the value of the seed is to take the milliseconds or similar rapidly changing set of digits from the real time clock of the computer. However, in cases when we wish to repeat a calculation using the same series of pseudo-random values, we can use `set.seed()` with an arbitrary integer as an argument to reset the generator to the same point in the underlying (deterministic) sequence.

📊📖 Execute the statement `rnorm(3)` by itself several times, paying attention to the values obtained. Repeat the exercise, but now executing `set.seed(98765)` immediately before each call to `rnorm(3)`, again paying attention to the values obtained. Next execute `set.seed(98765)`, followed by `c(rnorm(3), rnorm(3))`, and then execute `set.seed(98765)`, followed by `rnorm(6)` and compare the output. Repeat the exercise using a different argument in the call to `set.seed()`. analyze the results and explain how `setseed()` affects the generation of pseudo-random numbers in R.

4.4 "Random" sampling

In addition to drawing values from a theoretical distribution, we can draw values from an existing set or collection of values. We call this operation (pseudo-)random sampling. The draws can be done either with replacement or without replacement. In the second case, all draws are taken from the whole set of values, making it possible for a given value to be drawn more than once. In the default case of not using replacement, subsequent draws are taken from the values remaining after removing the values chosen in earlier draws.

```
sample(x = LETTERS)
##  [1] "Z" "N" "Y" "R" "M" "E" "W" "J" "H" "G" "U" "O" "S" "T" "L" "F" "X" "P" "K"
## [20] "V" "D" "A" "B" "C" "I" "Q"

sample(x = LETTERS, size = 12)
##  [1] "M" "S" "L" "R" "B" "D" "Q" "W" "V" "N" "J" "P"

sample(x = LETTERS, size = 12, replace = TRUE)
##  [1] "K" "E" "V" "N" "A" "Q" "L" "C" "T" "L" "H" "U"
```

In practice, pseudo-random sampling is useful when we need to select subsets of observations. One such case is assigning treatments to experimental units in an experiment or selecting persons to interview in a survey. Another use is in bootstrapping to estimate variation in parameter estimates using empirical distributions.

4.5 Correlation

Both parametric (Pearson's) and non-parametric robust (Spearman's and Kendall's) methods for the estimation of the (linear) correlation between pairs of variables are available in base R. The different methods are selected by passing arguments to a single function. While Pearson's method is based on the actual values of the observations, non-parametric methods are based on the ordering or rank of the observations, and consequently less affected by observations with extreme values.

We first load and explore the data set `cars` from R which we will use in the example. These data consist of stopping distances for cars moving at different speeds as described in the documentation available by entering `help(cars)`).

```
data(cars)
plot(cars)
```

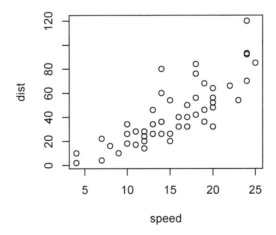

4.5.1 Pearson's r

Function `cor()` can be called with two vectors of the same length as arguments. In the case of the parametric Pearson method, we do not need to provide further arguments as this method is the default one.

```
cor(x = cars$speed, y = cars$dist)
## [1] 0.8068949
```

It is also possible to pass a data frame (or a matrix) as the only argument.

When the data frame (or matrix) contains only two columns, the returned value is equivalent to that of passing the two columns individually as vectors.

```
cor(cars)
##            speed      dist
## speed 1.0000000 0.8068949
## dist  0.8068949 1.0000000
```

When the data frame or matrix contains more than two numeric vectors, the returned value is a matrix of estimates of pairwise correlations between columns. We here use `rnorm()` described above to create a long vector of pseudo-random values drawn from the Normal distribution and `matrix()` to convert it into a matrix with three columns (see page 51 for details about R matrices).

```
my.mat <- matrix(rnorm(54), ncol = 3,
                 dimnames = list(rows = 1:18, cols = c("A", "B", "C")))
cor(my.mat)
##            A         B          C
## A 1.00000000 0.2126595 0.05623007
## B 0.21265951 1.0000000 0.31065243
## C 0.05623007 0.3106524 1.00000000
```

> 📘 Modify the code in the chunk immediately above constructing a matrix with six columns and then computing the correlations.

While `cor()` returns and estimate for r the correlation coefficient, `cor.test()` also computes the t-value, p-value, and confidence interval for the estimate.

```
cor.test(x = cars$speed, y = cars$dist)
##
##  Pearson's product-moment correlation
##
## data:  cars$speed and cars$dist
## t = 9.464, df = 48, p-value = 1.49e-12
## alternative hypothesis: true correlation is not equal to 0
## 95 percent confidence interval:
##  0.6816422 0.8862036
## sample estimates:
##       cor
## 0.8068949
```

As described below for model fitting and t-test, `cor.test()` also accepts a `formula` plus `data` as arguments.

> 📘 Functions `cor()` and `cor.test()` return R objects, that when using R interactively get automatically "printed" on the screen. One should be aware that `print()` methods do not necessarily display all the information contained in an R object. This is almost always the case for complex objects like those returned by R functions implementing statistical tests. As with any R object we can save

the result of an analysis into a variable. As described in section 2.13 on page 62 for lists, we can peek into the structure of an object with method str(). We can use class() and attributes() to extract further information. Run the code in the chunk below to discover what is actually returned by cor().

```r
a <- cor(cars)
class(a)
attributes(a)
str(a)
```

Methods class(), attributes() and str() are very powerful tools that can be used when we are in doubt about the data contained in an object and/or how it is structured. Knowing the structure allows us to retrieve the data members directly from the object when predefined extractor methods are not available.

4.5.2   Kendall's τ and Spearman's ρ

We use the same functions as for Pearson's r but explicitly request the use of one of these methods by passing and argument.

```r
cor(x = cars$speed, y = cars$dist, method = "kendall")
## [1] 0.6689901

cor(x = cars$speed, y = cars$dist, method = "spearman")
## [1] 0.8303568
```

Function cor.test(), described above, also allows the choice of method with the same syntax as shown for cor().

Repeat the exercise in the playground immediately above, but now using non-parametric methods. How does the information stored in the returned matrix differ depending on the method, and how can we extract information about the method used for calculation of the correlation from the returned object.

4.6   Fitting linear models

In R, the models to be fitted are described by "model formulas" such as y ~ x which we read as y is explained by x. Model formulas are used in different contexts: fitting of models, plotting, and tests like t-test. The syntax of model formulas is consistent throughout base R and numerous independently developed packages. However, their use is not universal, and several packages extend the basic syntax to allow the description of specific types of models.

As most things in R, model formulas can be stored in variables. In addition, contrary to the usual behavior of other statistical software, the result of a model fit is returned as an object, containing the different components of the fit. Once the model has been fitted, different methods allow us to extract parts and/or further manipulate the results obtained by fitting a model. Most of these methods have implementations for model fit objects for different types of statistical models. Consequently, what is described in this chapter using linear models as examples, also applies in many respects to the fit of models not described here.

The R function lm() is used to fit linear models. If the explanatory variable is continuous, the fit is a regression. If the explanatory variable is a factor, the fit is an analysis of variance (ANOVA) in broad terms. However, there is another meaning of ANOVA, referring only to the tests of significance rather to an approach to model fitting. Consequently, rather confusingly, results for tests of significance for fitted parameter estimates can both in the case of regression and ANOVA, be presented in an ANOVA table. In this second, stricter meaning, ANOVA means a test of significance based on the ratios between pairs of variances.

> ⚠ If you do not clearly remember the difference between numeric vectors and factors, or how they can be created, please, revisit chapter 2 on page 17.

4.6.1 Regression

In the example immediately below, speed is a continuous numeric variable. In the ANOVA table calculated for the model fit, in this case a linear regression, we can see that the term for speed has only one degree of freedom (df).

In the next example we continue using the stopping distance for cars data set included in R. Please see the plot on page 125.

```
data(cars)
is.factor(cars$speed)
## [1] FALSE

is.numeric(cars$speed)
## [1] TRUE
```

We then fit the simple linear model $y = \alpha \cdot 1 + \beta \cdot x$ where y corresponds to stopping distance (dist) and x to initial speed (speed). Such a model is formulated in R as dist ~ 1 + speed. We save the fitted model as fm1 (a mnemonic for fitted-model one).

```
fm1 <- lm(dist ~ 1 + speed, data=cars)
class(fm1)
## [1] "lm"
```

The next step is diagnosis of the fit. Are assumptions of the linear model procedure used reasonably close to being fulfilled? In R it is most common to use plots to this end. We show here only one of the four plots normally produced. This quantile vs. quantile plot allows us to assess how much the residuals deviate from being normally distributed.

```
plot(fm1, which = 2)
```

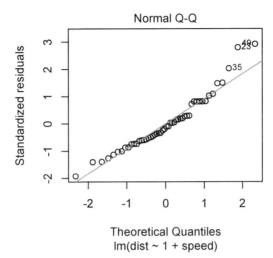

In the case of a regression, calling `summary()` with the fitted model object as argument is most useful as it provides a table of coefficient estimates and their errors. Remember that as is the case for most R functions, the value returned by `summary()` is printed when we call this method at the R prompt.

```
summary(fm1)
##
## Call:
## lm(formula = dist ~ 1 + speed, data = cars)
##
## Residuals:
##     Min      1Q  Median      3Q     Max
## -29.069  -9.525  -2.272   9.215  43.201
##
## Coefficients:
##             Estimate Std. Error t value Pr(>|t|)
## (Intercept) -17.5791     6.7584  -2.601   0.0123 *
## speed         3.9324     0.4155   9.464 1.49e-12 ***
## ---
## Signif. codes:  0 '***' 0.001 '**' 0.01 '*' 0.05 '.' 0.1 ' ' 1
##
## Residual standard error: 15.38 on 48 degrees of freedom
## Multiple R-squared:  0.6511,Adjusted R-squared:  0.6438
## F-statistic: 89.57 on 1 and 48 DF,  p-value: 1.49e-12
```

Let's look at the printout of the summary, section by section. Under "Call:" we find, `dist ~ 1 + speed` or the specification of the model fitted, plus the data used. Under "Residuals:" we find the extremes, quartiles and median of the residuals, or deviations between observations and the fitted line. Under "Coefficients:" we find the estimates of the model parameters and their variation plus corresponding *t*-tests. At the end of the summary there is information on degrees of freedom and overall coefficient of determination (R^2).

If we return to the model formulation, we can now replace α and β by the estimates obtaining $y = -17.6 + 3.93x$. Given the nature of the problem, we *know based on first principles* that stopping distance must be zero when speed is zero. This suggests that we should not estimate the value of α but instead set $\alpha = 0$, or in other words, fit the model $y = \beta \cdot x$.

However, in R models, the intercept is always implicitly included, so the model fitted above can be formulated as dist ~ speed—i.e., a missing + 1 does not change the model. To exclude the intercept from the previous model, we need to specify it as dist ~ speed - 1, resulting in the fitting of a straight line passing through the origin ($x = 0$, $y = 0$).

```
fm2 <- lm(dist ~ speed - 1, data = cars)
summary(fm2)
##
## Call:
## lm(formula = dist ~ speed - 1, data = cars)
##
## Residuals:
##     Min      1Q  Median      3Q     Max
## -26.183 -12.637  -5.455   4.590  50.181
##
## Coefficients:
##       Estimate Std. Error t value Pr(>|t|)
## speed   2.9091     0.1414   20.58   <2e-16 ***
## ---
## Signif. codes:  0 '***' 0.001 '**' 0.01 '*' 0.05 '.' 0.1 ' ' 1
##
## Residual standard error: 16.26 on 49 degrees of freedom
## Multiple R-squared:  0.8963,Adjusted R-squared:  0.8942
## F-statistic: 423.5 on 1 and 49 DF,  p-value: < 2.2e-16
```

Now there is no estimate for the intercept in the summary, only an estimate for the slope.

```
plot(fm2, which = 1)
```

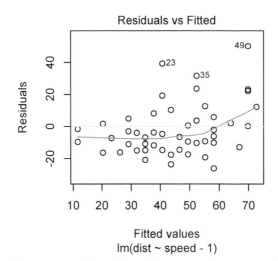

The equation of the second fitted model is $y = 2.91x$, and from the residuals, it can be seen that it is inadequate, as the straight line does not follow the curvature of the relationship between dist and speed.

![icon] You will now fit a second-degree polynomial, a different linear model: $y = \alpha \cdot 1 + \beta_1 \cdot x + \beta_2 \cdot x^2$. The function used is the same as for linear regression, lm(). We only need to alter the formulation of the model. The identity function I() is used to protect its argument from being interpreted as part of the model formula. Instead, its argument is evaluated beforehand and the result is used as the, in this case second, explanatory variable.

```
fm3 <- lm(dist ~ speed + I(speed^2), data = cars)
plot(fm3, which = 3)
summary(fm3)
anova(fm3)
```

The "same" fit using an orthogonal polynomial can be specified using function poly(). Polynomials of different degrees can be obtained by supplying as the second argument to poly() the corresponding positive integer value. In this case, the different terms of the polynomial are bulked together in the summary.

```
fm3a <- lm(dist ~ poly(speed, 2), data = cars)
summary(fm3a)
anova(fm3a)
```

We can also compare two model fits using anova(), to test whether one of the models describes the data better than the other. It is important in this case to take into consideration the nature of the difference between the model formulas, most importantly if they can be interpreted as nested—i.e., interpreted as a base model vs. the same model with additional terms.

```
anova(fm2, fm1)
```

Three or more models can also be compared in a single call to anova(). However, be careful, as the order of the arguments matters.

```
anova(fm2, fm3, fm3a)
anova(fm2, fm3a, fm3)
```

We can use different criteria to choose the "best" model: significance based on *p*-values or information criteria (AIC, BIC). AIC (Akaike's "An Information Criterion") and BIC ("Bayesian Information Criterion" = SBC, "Schwarz's Bayesian criterion") that penalize the resulting "goodness" based on the number of parameters in the fitted model. In the case of AIC and BIC, a smaller value is better, and values returned can be either positive or negative, in which case more negative is better. Estimates for both BIC and AIC are returned by anova(), and on their own by BIC() and AIC()

```
BIC(fm2, fm1, fm3, fm3a)
AIC(fm2, fm1, fm3, fm3a)
```

Once you have run the code in the chunks above, you will be able see that these three criteria do not necessarily agree on which is the "best" model. Find in the output p-value, BIC and AIC estimates, for the different models and conclude which model is favored by each of the three criteria. In addition you will notice that the two different formulations of the quadratic polynomial are equivalent.

Additional methods give easy access to different components of fitted models: vcov() returns the variance-covariance matrix, coef() and its alias coefficients() return the estimates for the fitted model coefficients, fitted() and its alias fitted.values() extract the fitted values, and resid() and its alias residuals() the corresponding residuals (or deviations). Less frequently used accessors are effects(), terms(), model.frame() and model.matrix().

Familiarize yourself with these extraction and summary methods by reading their documentation and use them to explore fm1 fitted above or model fits to other data of your interest.

The objects returned by model fitting functions are rather complex and contain the full information, including the data to which the model was fit to. The different functions described above, either extract parts of the object or do additional calculations and formatting based on them. There are different specializations of these methods which are called depending on the class of the model-fit object. (See section 5.4 on page 172.)

```
class(fm1)
## [1] "lm"
```

We rarely need to manually explore the structure of these model-fit objects when using R interactively. In contrast, when including model fitting in scripts or package code, the need to efficiently extract specific members from them happens more frequently. As with any other R object we can use str() to explore them.

```
str(fm1, max.level = 1) # not evaluated
```

We frequently only look at the output of anova() as implicitly displayed by print(). However, both anova() and summary() return complex objects containing members with data not displayed by the matching print() methods. Understanding this is frequently useful, when we want to either display the results in a different format, or extract parts of them for use in additional tests or computations. Once again we use str() to look at the structure.

```
str(anova(fm1))
## Classes 'anova' and 'data.frame': 2 obs. of  5 variables:
##  $ Df     : int  1 48
##  $ Sum Sq : num  21185 11354
##  $ Mean Sq: num  21185 237
##  $ F value: num  89.6 NA
##  $ Pr(>F) : num  1.49e-12 NA
##  - attr(*, "heading")= chr [1:2] "Analysis of Variance Table\n" "Response: dist"


str(summary(fm1))
## List of 11
##  $ call         : language lm(formula = dist ~ 1 + speed, data = cars)
##  $ terms        :Classes 'terms', 'formula'  language dist ~ 1 + speed
##   .. ..- attr(*, "variables")= language list(dist, speed)
##   .. ..- attr(*, "factors")= int [1:2, 1] 0 1
##   .. .. ..- attr(*, "dimnames")=List of 2
##   .. .. .. ..$ : chr [1:2] "dist" "speed"
##   .. .. .. ..$ : chr "speed"
##   .. ..- attr(*, "term.labels")= chr "speed"
##   .. ..- attr(*, "order")= int 1
##   .. ..- attr(*, "intercept")= int 1
##   .. ..- attr(*, "response")= int 1
##   .. ..- attr(*, ".Environment")=<environment: R_GlobalEnv>
##   .. ..- attr(*, "predvars")= language list(dist, speed)
##   .. ..- attr(*, "dataClasses")= Named chr [1:2] "numeric" "numeric"
##   .. .. ..- attr(*, "names")= chr [1:2] "dist" "speed"
##  $ residuals    : Named num [1:50] 3.85 11.85 -5.95 12.05 2.12 ...
##   ..- attr(*, "names")= chr [1:50] "1" "2" "3" "4" ...
##  $ coefficients : num [1:2, 1:4] -17.579 3.932 6.758 0.416 -2.601 ...
##   ..- attr(*, "dimnames")=List of 2
##   .. ..$ : chr [1:2] "(Intercept)" "speed"
##   .. ..$ : chr [1:4] "Estimate" "Std. Error" "t value" "Pr(>|t|)"
##  $ aliased      : Named logi [1:2] FALSE FALSE
##   ..- attr(*, "names")= chr [1:2] "(Intercept)" "speed"
##  $ sigma        : num 15.4
##  $ df           : int [1:3] 2 48 2
##  $ r.squared    : num 0.651
##  $ adj.r.squared: num 0.644
##  $ fstatistic   : Named num [1:3] 89.6 1 48
##   ..- attr(*, "names")= chr [1:3] "value" "numdf" "dendf"
##  $ cov.unscaled : num [1:2, 1:2] 0.19311 -0.01124 -0.01124 0.00073
##   ..- attr(*, "dimnames")=List of 2
##   .. ..$ : chr [1:2] "(Intercept)" "speed"
##   .. ..$ : chr [1:2] "(Intercept)" "speed"
##  - attr(*, "class")= chr "summary.lm"
```

Once we know the structure of the object and the names of members, we can simply extract them using the usual R rules for member extraction.

```
summary(fm1)$adj.r.squared
## [1] 0.6438102
```

As an example we test if the slope from a linear regression fit deviates significantly from a constant value different from the usual zero.

The examples above are for a null hypothesis of slope = 0 and next we show how to do the equivalent test with a null hypothesis of slope = 1. The procedure is applicable to any constant value as a null hypothesis for any of the fitted parameter estimates for hypotheses set *a priori*. The examples use a two-sided test. In some cases, a single-sided test should be used (e.g., if its known a priori that deviation is because of physical reasons possible only in one direction away from the null hypothesis, or because only one direction of response is of interest).

To estimate the *t*-value we need an estimate for the parameter and an estimate of the standard error for this estimate and its degrees of freedom.

```
est.slope.value <- summary(fm1)$coef["speed", "Estimate"]
est.slope.se <- summary(fm1)$coef["speed", "Std. Error"]
degrees.of.freedom <- summary(fm1)$df[2]
```

The *t*-test is based on the difference between the value of the null hypothesis and the estimate.

```
hyp.null <- 1
t.value <- (est.slope.value - hyp.null) / est.slope.se
p.value <- dt(t.value, df = degrees.of.freedom)
```

Check that the procedure above agrees with the output of `summary()` when we set `hyp.null <- 0` instead of `hyp.null <- 1`.

Modify the example so as to test whether the intercept is significantly larger than 5 feet, doing a one-sided test.

Method `predict()` uses the fitted model together with new data for the independent variables to compute predictions. As `predict()` accepts new data as input, it allows interpolation and extrapolation to values of the independent variables not present in the original data. In the case of fits of linear- and some other models, method `predict()` returns, in addition to the prediction, estimates of the confidence and/or prediction intervals. The new data must be stored in a data frame with columns using the same names for the explanatory variables as in the data used for the fit, a response variable is not needed and additional columns are ignored. (The explanatory variables in the new data can be either continuous or factors, but they must match in this respect those in the original data.)

Predict using both `fm1` and `fm2` the distance required to stop cars moving at 0, 5, 10, 20, 30, and 40 mph. Study the help page for the predict method for linear models (using `help(predict.lm)`). Explore the difference between `"prediction"` and `"confidence"` bands: why are they so different?

4.6.2 Analysis of variance, ANOVA

We use here the `InsectSprays` data set, giving insect counts in plots sprayed with different insecticides. In these data, `spray` is a factor with six levels.

The call is exactly the same as the one for linear regression, only the names of the variables and data frame are different. What determines that this is an ANOVA is that `spray`, the explanatory variable, is a `factor`.

```
data(InsectSprays)
is.numeric(InsectSprays$spray)
## [1] FALSE

is.factor(InsectSprays$spray)
## [1] TRUE

levels(InsectSprays$spray)
## [1] "A" "B" "C" "D" "E" "F"
```

We fit the model in exactly the same way as for linear regression; the difference is that we use a factor as the explanatory variable. By using a factor instead of a numeric vector, a different model matrix is built from an equivalent formula.

```
fm4 <- lm(count ~ spray, data = InsectSprays)
```

Diagnostic plots are obtained in the same way as for linear regression.

```
plot(fm4, which = 3)
```

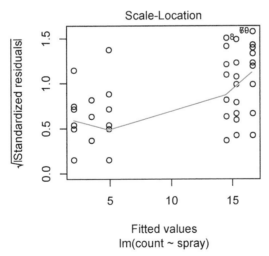

In ANOVA we are mainly interested in testing hypotheses, and `anova()` provides the most interesting output. Function `summary()` can be used to extract parameter estimates. The default contrasts and corresponding p-values returned by `summary()` test hypotheses that have little or no direct interest in an analysis of variance. Function `aov()` is a wrapper on `lm()` that returns an object that by default when printed displays the output of `anova()`.

```
anova(fm4)
## Analysis of Variance Table
##
## Response: count
##           Df Sum Sq Mean Sq F value    Pr(>F)
## spray      5 2668.8  533.77  34.702 < 2.2e-16 ***
## Residuals 66 1015.2   15.38
## ---
## Signif. codes:  0 '***' 0.001 '**' 0.01 '*' 0.05 '.' 0.1 ' ' 1
```

⚠ The defaults used for model fits and ANOVA calculations vary among programs. There exist different so-called "types" of sums of squares, usually called I, II, and III. In orthogonal designs the choice has no consequences, but differences can be important for unbalanced designs, even leading to different conclusions. R's default, type I, is usually considered to suffer milder problems than type III, the default used by SPSS and SAS.

The contrasts used affect the estimates returned by `coef()` and `summary()` applied to an ANOVA model fit. The default used in R is different to that used in some other programs (even different than in S). The most straightforward way of setting a different default for a whole series of model fits is by setting R option `contrasts`, which we here only print.

```
options("contrasts")
## $contrasts
##          unordered          ordered
## "contr.treatment"      "contr.poly"
```

It is also possible to select the contrast to be used in the call to `aov()` or `lm()`. The default, `contr.treatment` uses the first level of the factor (assumed to be a control) as reference for estimation of coefficients and their significance, while `contr.sum` uses as reference the mean of all levels, by using as condition that the sum of the coefficient estimates is equal to zero. Obviously this changes what the coefficients describe, and consequently also the estimated *p*-values.

```
fm4trea <- lm(count ~ spray, data = InsectSprays,
              contrasts = list(spray = contr.treatment))
fm4sum  <- lm(count ~ spray, data = InsectSprays,
              contrasts = list(spray = contr.sum))
```

Interpretation of any analysis has to take into account these differences and users should not be surprised if ANOVA yields different results in base R and SPSS or SAS given the different types of sums of squares used. The interpretation of ANOVA on designs that are not orthogonal will depend on which type is used, so the different results are not necessarily contradictory even when different.

```
summary(fm4trea)
##
## Call:
## lm(formula = count ~ spray, data = InsectSprays,
##     contrasts = list(spray = contr.treatment))
##
## Residuals:
##    Min     1Q Median    3Q    Max
## -8.333 -1.958 -0.500  1.667  9.333
##
## Coefficients:
##             Estimate Std. Error t value Pr(>|t|)
## (Intercept)  14.5000     1.1322  12.807  < 2e-16 ***
## spray2        0.8333     1.6011   0.520    0.604
## spray3      -12.4167     1.6011  -7.755 7.27e-11 ***
## spray4       -9.5833     1.6011  -5.985 9.82e-08 ***
## spray5      -11.0000     1.6011  -6.870 2.75e-09 ***
## spray6        2.1667     1.6011   1.353    0.181
## ---
## Signif. codes:  0 '***' 0.001 '**' 0.01 '*' 0.05 '.' 0.1 ' ' 1
##
## Residual standard error: 3.922 on 66 degrees of freedom
## Multiple R-squared:  0.7244,Adjusted R-squared:  0.7036
## F-statistic:  34.7 on 5 and 66 DF,  p-value: < 2.2e-16
```

```
summary(fm4sum)
##
## Call:
## lm(formula = count ~ spray, data = InsectSprays,
##     contrasts = list(spray = contr.sum))
##
## Residuals:
##    Min     1Q Median    3Q    Max
## -8.333 -1.958 -0.500  1.667  9.333
##
## Coefficients:
##             Estimate Std. Error t value Pr(>|t|)
## (Intercept)   9.5000     0.4622  20.554  < 2e-16 ***
## spray1        5.0000     1.0335   4.838 8.22e-06 ***
## spray2        5.8333     1.0335   5.644 3.78e-07 ***
## spray3       -7.4167     1.0335  -7.176 7.87e-10 ***
## spray4       -4.5833     1.0335  -4.435 3.57e-05 ***
## spray5       -6.0000     1.0335  -5.805 2.00e-07 ***
## ---
## Signif. codes:  0 '***' 0.001 '**' 0.01 '*' 0.05 '.' 0.1 ' ' 1
##
## Residual standard error: 3.922 on 66 degrees of freedom
## Multiple R-squared:  0.7244,Adjusted R-squared:  0.7036
## F-statistic:  34.7 on 5 and 66 DF,  p-value: < 2.2e-16
```

In the case of contrasts, they always affect the parameter estimates independently of whether the experiment design is orthogonal or not. A different set of contrasts simply tests a different set of possible treatment effects. Contrasts, on the other hand, do not affect the table returned by anova() as this table does not deal with the effects of individual factor levels.

4.6.3   Analysis of covariance, ANCOVA

When a linear model includes both explanatory factors and continuous explanatory variables, we may call it *analysis of covariance* (ANCOVA). The formula syntax is the same for all linear models and, as mentioned in previous sections, what determines the type of analysis is the nature of the explanatory variable(s). As the formulation remains the same, no specific example is given. The main difficulty of ANCOVA is in the selection of the covariate and the interpretation of the results of the analysis (e.g. Smith 1957).

4.7   Generalized linear models

Linear models make the assumption of normally distributed residuals. Generalized linear models, fitted with function `glm()` are more flexible, and allow the assumed distribution to be selected as well as the link function. For the analysis of the `InsectSpray` data set above (section 4.6.2 on page 135), the Normal distribution is not a good approximation as count data deviates from it. This was visible in the quantile–quantile plot above.

For count data, GLMs provide a better alternative. In the example below we fit the same model as above, but we assume a quasi-Poisson distribution instead of the Normal. In addition to the model formula we need to pass an argument through `family` giving the error distribution to be assumed—the default for `family` is `gaussian` or Normal distribution.

```
fm10 <- glm(count ~ spray, data = InsectSprays, family = quasipoisson)
anova(fm10)
## Analysis of Deviance Table
##
## Model: quasipoisson, link: log
##
## Response: count
##
## Terms added sequentially (first to last)
##
##
##       Df Deviance Resid. Df Resid. Dev
## NULL                     71     409.04
## spray  5   310.71        66      98.33
```

The printout from the `anova()` method for GLM fits has some differences to that for LM fits. By default, no significance test is computed, as a knowledgeable choice is required depending on the characteristics of the model and data. We here use `"F"` as an argument to request an *F*-test.

```
anova(fm10, test = "F")
## Analysis of Deviance Table
##
## Model: quasipoisson, link: log
##
## Response: count
```

```
##
## Terms added sequentially (first to last)
##
##
##        Df Deviance Resid. Df Resid. Dev      F    Pr(>F)
## NULL                     71     409.04
## spray  5   310.71        66      98.33 41.216 < 2.2e-16 ***
## ---
## Signif. codes:  0 '***' 0.001 '**' 0.01 '*' 0.05 '.' 0.1 ' ' 1
```

Method `plot()` as for linear-model fits, produces diagnosis plots. We show as above the q-q-plot of residuals.

```
plot(fm10, which = 3)
```

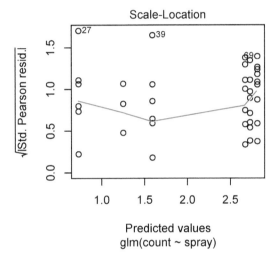

We can extract different components similarly as described for linear models (see section 4.6 on page 127).

```
class(fm10)
## [1] "glm" "lm"

summary(fm10)
##
## Call:
## glm(formula = count ~ spray, family = quasipoisson, data = InsectSprays)
##
## Deviance Residuals:
##     Min      1Q   Median      3Q     Max
## -2.3852  -0.8876  -0.1482  0.6063  2.6922
##
## Coefficients:
##             Estimate Std. Error t value Pr(>|t|)
## (Intercept)  2.67415    0.09309  28.728  < 2e-16 ***
## sprayB       0.05588    0.12984   0.430    0.668
## sprayC      -1.94018    0.26263  -7.388 3.30e-10 ***
## sprayD      -1.08152    0.18499  -5.847 1.70e-07 ***
## sprayE      -1.42139    0.21110  -6.733 4.82e-09 ***
```

```
## sprayF        0.13926    0.12729    1.094     0.278
## ---
## Signif. codes:  0 '***' 0.001 '**' 0.01 '*' 0.05 '.' 0.1 ' ' 1
##
## (Dispersion parameter for quasipoisson family taken to be 1.507713)
##
##     Null deviance: 409.041  on 71  degrees of freedom
## Residual deviance:  98.329  on 66  degrees of freedom
## AIC: NA
##
## Number of Fisher Scoring iterations: 5

head(residuals(fm10))
##         1          2          3          4          5          6
## -1.2524891 -2.1919537  1.3650439 -0.1320721 -0.1320721 -0.6768988

head(fitted(fm10))
##    1    2    3    4    5    6
## 14.5 14.5 14.5 14.5 14.5 14.5
```

☕ If we use `str()` or `names()` we can see that there are some differences with respect to linear model fits. The returned object is of a different class and contains some members not present in linear models. Two of these have to do with the iterative approximation method used, `iter` contains the number of iterations used and `converged` the success or not in finding a solution.

```
names(fm10)
##  [1] "coefficients"    "residuals"       "fitted.values"
##  [4] "effects"         "R"               "rank"
##  [7] "qr"              "family"          "linear.predictors"
## [10] "deviance"        "aic"             "null.deviance"
## [13] "iter"            "weights"         "prior.weights"
## [16] "df.residual"     "df.null"         "y"
## [19] "converged"       "boundary"        "model"
## [22] "call"            "formula"         "terms"
## [25] "data"            "offset"          "control"
## [28] "method"          "contrasts"       "xlevels"

fm10$converged
## [1] TRUE

fm10$iter
## [1] 5
```

4.8   Non-linear regression

Function `nls()` is R's workhorse for fitting non-linear models. By *non-linear* it is meant non-linear *in the parameters* whose values are being estimated through fitting the model to data. This is different from the shape of the function when plotted—i.e., polynomials of any degree are linear models. In contrast, the Michaelis-Menten equation used in chemistry and the Gompertz equation used to describe growth are non-linear models in their parameters.

While analytical algorithms exist for finding estimates for the parameters of linear models, in the case of non-linear models, the estimates are obtained by approximation. For analytical solutions, estimates can always be obtained, except in infrequent pathological cases where reliance on floating point numbers with limited resolution introduces rounding errors that "break" mathematical algorithms that are valid for real numbers. For approximations obtained through iteration, cases when the algorithm fails to *converge* onto an answer are relatively common. Iterative algorithms attempt to improve an initial guess for the values of the parameters to be estimated, a guess frequently supplied by the user. In each iteration the estimate obtained in the previous iteration is used as the starting value, and this process is repeated one time after another. The expectation is that after a finite number of iterations the algorithm will converge into a solution that "cannot" be improved further. In real life we stop iteration when the improvement in the fit is smaller than a certain threshold, or when no convergence has been achieved after a certain maximum number of iterations. In the first case, we usually obtain good estimates; in the second case, we do not obtain usable estimates and need to look for different ways of obtaining them. When convergence fails, the first thing to do is to try different starting values and if this also fails, switch to a different computational algorithm. These steps usually help, but not always. Good starting values are in many cases crucial and in some cases "guesses" can be obtained using either graphical or analytical approximations.

For functions for which computational algorithms exist for "guessing" suitable starting values, R provides a mechanism for packaging the function to be fitted together with the function generating the starting values. These functions go by the name of *self-starting functions* and relieve the user from the burden of guessing and supplying suitable starting values. The self-starting functions available in R are `SSasymp()`, `SSasympOff()`, `SSasympOrig()`, `SSbiexp()`, `SSfol()`, `SSfpl()`, `SSgompertz()`, `SSlogis()`, `SSmicmen()`, and `SSweibull()`. Function `selfStart()` can be used to define new ones. All these functions can be used when fitting models with `nls` or `nlme`. Please, check the respective help pages for details.

In the case of `nls()` the specification of the model to be fitted differs from that used for linear models. We will use as an example fitting the Michaelis-Menten equation describing reaction kinetics in biochemistry and chemistry. The mathematical formulation is given by:

$$v = \frac{\mathrm{d}[P]}{\mathrm{d}t} = \frac{V_{\max}[S]}{K_{\mathrm{M}} + [S]} \tag{4.1}$$

The function takes its name from Michaelis and Menten's paper from 1913 (Johnson and Goody 2011). A self-starting function implementing the Michaelis-Menten equation is available in R under the name `SSmicmen()` . We will use the `Puromycin` data set.

```
data(Puromycin)
names(Puromycin)
## [1] "conc"  "rate"  "state"
```

```
fm21 <- nls(rate ~ SSmicmen(conc, Vm, K), data = Puromycin,
            subset = state == "treated")
```

We can extract different components similarly as described for linear models (see section 4.6 on page 127).

```
class(fm21)
## [1] "nls"
```

```
summary(fm21)
##
## Formula: rate ~ SSmicmen(conc, Vm, K)
##
## Parameters:
##     Estimate Std. Error t value Pr(>|t|)
## Vm 2.127e+02  6.947e+00   30.615 3.24e-11 ***
## K  6.412e-02  8.281e-03    7.743 1.57e-05 ***
## ---
## Signif. codes:  0 '***' 0.001 '**' 0.01 '*' 0.05 '.' 0.1 ' ' 1
##
## Residual standard error: 10.93 on 10 degrees of freedom
##
## Number of iterations to convergence: 0
## Achieved convergence tolerance: 1.937e-06
```

```
residuals(fm21)
##   [1]  25.4339970  -3.5660030  -5.8109606   4.1890394 -11.3616076   4.6383924
##   [7]  -5.6846886 -12.6846886   0.1670799  10.1670799   6.0311724  -0.9688276
## attr(,"label")
## [1] "Residuals"
```

```
fitted(fm21)
##   [1]  50.5660  50.5660 102.8110 102.8110 134.3616 134.3616 164.6847 164.6847
##   [9] 190.8329 190.8329 200.9688 200.9688
## attr(,"label")
## [1] "Fitted values"
```

> 💻 If we use str() or names() we can see that there are differences with respect to linear model and generalized model fits. The returned object is of class nls and contains some new members and lacks others. Two members are related to the iterative approximation method used, control containing nested members holding iteration settings, and convInfo (convergence information) with nested members with information on the outcome of the iterative algorithm.

```
str(fm21, max.level = 1)
## List of 6
##  $ m            :List of 16
##   ..- attr(*, "class")= chr "nlsModel"
##  $ convInfo   :List of 5
##  $ data       : symbol Puromycin
##  $ call       : language nls(formula = rate ~ SSmicmen(conc, Vm, K),
##                             data = Puromycin, subset = __truncated__ ...
##  $ dataClasses: Named chr "numeric"
##   ..- attr(*, "names")= chr "conc"
##  $ control    :List of 5
##  - attr(*, "class")= chr "nls"


fm21$convInfo
## $isConv
## [1] TRUE
##
## $finIter
## [1] 0
##
## $finTol
## [1] 1.937028e-06
##
## $stopCode
## [1] 0
##
## $stopMessage
## [1] "converged"
```

4.9   Model formulas

In the examples above we fitted simple models. More complex ones can be easily formulated using the same syntax. First of all, one can avoid use of operator * and explicitly define all individual main effects and interactions using operators + and :. The syntax implemented in base R allows grouping by means of parentheses, so it is also possible to exclude some interactions by combining the use of * and parentheses.

The same symbols as for arithmetic operators are used for model formulas. Within a formula, symbols are interpreted according to formula syntax. When we mean an arithmetic operation that could be interpreted as being part of the model formula we need to "protect" it by means of the identity function I(). The next two examples define formulas for models with only one explanatory variable. With formulas like these, the explanatory variable will be computed on the fly when fitting the model to data. In the first case below we need to explicitly protect the addition of the two variables into their sum, because otherwise they would be interpreted as two separate explanatory variables in the model. In the second case, log() cannot be interpreted as part of the model formula, and consequently does not require additional protection, neither does the expression passed as its argument.

```
y ~ I(x1 + x2)
y ~ log(x1 + x2)
```

R formula syntax allows alternative ways for specifying interaction terms. They allow "abbreviated" ways of entering formulas, which for complex experimental designs saves typing and can improve clarity. As seen above, operator * saves us from having to explicitly indicate all the interaction terms in a full factorial model.

```
y ~ x1 + x2 + x3 + x1:x2 + x1:x3 + x2:x3 + x1:x2:x3
```

Can be replaced by a concise equivalent.

```
y ~ x1 * x2 * x3
```

When the model to be specified does not include all possible interaction terms, we can combine the concise notation with parentheses.

```
y ~ x1 + (x2 * x3)
y ~ x1 + x2 + x3 + x2:x3
```

That the two model formulas above are equivalent, can be seen using terms()

```
terms(y ~ x1 + (x2 * x3))
## y ~ x1 + (x2 * x3)
## attr(,"variables")
## list(y, x1, x2, x3)
## attr(,"factors")
##     x1 x2 x3 x2:x3
## y   0  0  0     0
## x1  1  0  0     0
## x2  0  1  0     1
## x3  0  0  1     1
## attr(,"term.labels")
## [1] "x1"    "x2"    "x3"    "x2:x3"
## attr(,"order")
## [1] 1 1 1 2
## attr(,"intercept")
## [1] 1
## attr(,"response")
## [1] 1
## attr(,".Environment")
## <environment: R_GlobalEnv>


y ~ x1 * (x2 + x3)
y ~ x1 + x2 + x3 + x1:x2 + x1:x3


terms(y ~ x1 * (x2 + x3))
## y ~ x1 * (x2 + x3)
## attr(,"variables")
## list(y, x1, x2, x3)
## attr(,"factors")
```

```
##     x1 x2 x3 x1:x2 x1:x3
## y   0  0  0     0     0
## x1  1  0  0     1     1
## x2  0  1  0     1     0
## x3  0  0  1     0     1
## attr(,"term.labels")
## [1] "x1"    "x2"    "x3"    "x1:x2" "x1:x3"
## attr(,"order")
## [1] 1 1 1 2 2
## attr(,"intercept")
## [1] 1
## attr(,"response")
## [1] 1
## attr(,".Environment")
## <environment: R_GlobalEnv>
```

The ∧ operator provides a concise notation to limit the order of the interaction terms included in a formula.

```
y ~ (x1 + x2 + x3)∧2
y ~ x1 + x2 + x3 + x1:x2 + x1:x3 + x2:x3
```

```
terms(y ~ (x1 + x2 + x3)∧2)
## y ~ (x1 + x2 + x3)∧2
## attr(,"variables")
## list(y, x1, x2, x3)
## attr(,"factors")
##     x1 x2 x3 x1:x2 x1:x3 x2:x3
## y   0  0  0     0     0     0
## x1  1  0  0     1     1     0
## x2  0  1  0     1     0     1
## x3  0  0  1     0     1     1
## attr(,"term.labels")
## [1] "x1"    "x2"    "x3"    "x1:x2" "x1:x3" "x2:x3"
## attr(,"order")
## [1] 1 1 1 2 2 2
## attr(,"intercept")
## [1] 1
## attr(,"response")
## [1] 1
## attr(,".Environment")
## <environment: R_GlobalEnv>
```

> For operator ∧ to behave as expected, its first operand should be a formula with no interactions! Compare the result of expanding these two formulas with `trems()`.
>
> ```
> y ~ (x1 + x2 + x3)∧2
> y ~ (x1 * x2 * x3)∧2
> ```

Operator `%in%` can also be used as a shortcut for including only some of all the possible interaction terms in a formula.

```
y ~ x1 + x2 + x1 %in% x2


terms(y ~ x1 + x2 + x1 %in% x2)
## y ~ x1 + x2 + x1 %in% x2
## attr(,"variables")
## list(y, x1, x2)
## attr(,"factors")
##     x1 x2 x1:x2
## y   0  0     0
## x1  1  0     1
## x2  0  1     1
## attr(,"term.labels")
## [1] "x1"     "x2"     "x1:x2"
## attr(,"order")
## [1] 1 1 2
## attr(,"intercept")
## [1] 1
## attr(,"response")
## [1] 1
## attr(,".Environment")
## <environment: R_GlobalEnv>
```

Execute the examples below using the npk data set from R. They demonstrate the use of different model formulas in ANOVA. Use these examples plus your own variations on the same theme to build your understanding of the syntax of model formulas. Based on the terms displayed in the ANOVA tables, first work out what models are being fitted in each case. In a second step, write each of the models using a mathematical formulation. Finally, think how model choice may affect the conclusions from an analysis of variance.

```
data(npk)
anova(lm(yield ~ N * P * K, data = npk))
anova(lm(yield ~ (N + P + K)^2, data = npk))
anova(lm(yield ~ N + P + K + P %in% N + K %in% N, data = npk))
anova(lm(yield ~ N + P + K + N %in% P + K %in% P, data = npk))
```

Nesting of factors in experiments using hierarchical designs such as split-plots or repeated measures, results in the need to compute additional error terms, differing in their degrees of freedom. In such a design, different effects are tested based on different error terms. Whether nesting exists or not is a property of an experiment. It is decided as part of the design of the experiment based on the mechanics of treatment assignment to experimental units. In base-R model-formulas, nesting needs to be described by explicit definition of error terms by means of Error() within the formula. Nowadays, linear mixed-effects (LME) models are most frequently used with data from experiments and surveys using hierarchical designs, as implemented in packages 'nlme' and 'lme4'. These two packages use their own extensions to the model formula syntax to describe nesting and distinguishing fixed and random effects. Additive models have required other extensions, most of them specific to individual packages. These extensions fall outside the scope of this book.

⚠ R will accept any syntactically correct model formula, even when the results of the fit are not interpretable. It is *the responsibility of the user to ensure that models are meaningful.* The most common, and dangerous, mistake is specifying for factorial experiments, models that are missing lower-order interactions.

Fitting models like those below to data from an experiment based on a three-way factorial design should be avoided. In both cases simpler terms are missing, while higher-order interaction(s) that include the missing term are included in the model. Such models are not interpretable, as the variation from the missing term(s) ends being "disguised" within the remaining terms, distorting their apparent significance and parameter estimates.

```
y ~ A + B + A:B + A:C + B:C
y ~ A + B + C + A:B + A:C + A:B:C
```

In contrast to those above, the models below are interpretable, even if not "full" models (not including all possible interactions).

```
y ~ A + B + C + A:B + A:C + B:C
y ~ (A + B + C)^2
y ~ A + B + C + B:C
y ~ A + B * C
```

As seen in chapter 6, almost everything in the R language is an object that can be stored and manipulated. Model formulas are also objects, objects of class "formula".

```
class(y ~ x)
## [1] "formula"
```

```
a <- y ~ x
class(a)
## [1] "formula"
```

There is no method is.formula() in base R, but we can easily test the class of an object with inherits().

```
inherits(a, "formula")
## [1] TRUE
```

💻 **Manipulation of model formulas.** Because this is a book about the R language, it is pertinent to describe how formulas can be manipulated. Formulas, as any other R objects, can be saved in variables including lists. Why is this useful? For example, if we want to fit several different models to the same data, we can write a `for` loop that walks through a list of model formulas. Or we can write a function that accepts one or more formulas as arguments.

The use of `for` *loops* for iteration over a list of model formulas is described in section 3.6 on page 115.

```
my.data <- data.frame(x = 1:10, y = (1:10) / 2 + rnorm(10))
anovas <- list()
formulas <- list(a = y ~ x - 1, b = y ~ x, c = y ~ x + x^2)
for (formula in formulas) {
 anovas <- c(anovas, list(lm(formula, data = my.data)))
 }
str(anovas, max.level = 1)
## List of 3
##  $ :List of 12
##   ..- attr(*, "class")= chr "lm"
##  $ :List of 12
##   ..- attr(*, "class")= chr "lm"
##  $ :List of 12
##   ..- attr(*, "class")= chr "lm"
```

As could be expected, a conversion constructor is available with name `as.formula()`. It is useful when formulas are input interactively by the user or read from text files. With `as.formula()` we can convert a character string into a formula.

```
my.string <- "y ~ x"
lm(as.formula(my.string), data = my.data)
##
## Call:
## lm(formula = as.formula(my.string), data = my.data)
##
## Coefficients:
## (Intercept)            x
##      1.4059       0.2839
```

As there are many functions for the manipulation of character strings available in base R and through extension packages, it is straightforward to build model formulas programmatically as strings. We can use functions like `paste()` to assemble a formula as text, and then use `as.formula()` to convert it to an object of class `formula`, usable for fitting a model.

```
my.string <- paste("y", "x", sep = "~")
lm(as.formula(my.string), data = my.data)
##
## Call:
## lm(formula = as.formula(my.string), data = my.data)
##
## Coefficients:
## (Intercept)            x
##      1.4059       0.2839
```

For the reverse operation of converting a formula into a string, we have available methods `as.character()` and `format()`. The first of these methods returns a character vector containing the components of the formula as individual strings, while `format()` returns a single character string with the formula formatted for printing.

```
formatted.string <- format(y ~ x)
formatted.string
## [1] "y ~ x"

as.formula(formatted.string)
## y ~ x
```

It is also possible to *edit* formula objects with method `update()`. In the replacement formula, a dot can replace either the left-hand side (lhs) or the right-hand side (rhs) of the existing formula in the replacement formula. We can also remove terms as can be seen below. In some cases the dot corresponding to the lhs can be omitted, but including it makes the syntax clearer.

```
my.formula <- y ~ x1 + x2
update(my.formula, . ~ . + x3)
## y ~ x1 + x2 + x3

update(my.formula, . ~ . - x1)
## y ~ x2

update(my.formula, . ~ x3)
## y ~ x3

update(my.formula, z ~ .)
## z ~ x1 + x2

update(my.formula, . + z ~ .)
## y + z ~ x1 + x2
```

R provides high-level functions for model selection. Consequently many R users will rarely need to edit model formulas in their scripts. For example, step-wise model selection is possible with R method `step()`.

A matrix of dummy coefficients can be derived from a model formula, a type of contrast, and the data for the explanatory variables.

```
treats.df <- data.frame(A = rep(c("yes", "no"), c(4, 4)),
                        B = rep(c("white", "black"), 4))
treats.df
##      A     B
## 1 yes white
## 2 yes black
## 3 yes white
## 4 yes black
## 5  no white
## 6  no black
## 7  no white
## 8  no black
```

The default contrasts types currently in use.

```
options("contrasts")
## $contrasts
##          unordered              ordered
## "contr.treatment"        "contr.poly"
```

A model matrix for a model for a two-way factorial design with no interaction term:

```
model.matrix(~ A + B, treats.df)
##    (Intercept) Ayes Bwhite
## 1            1    1      1
## 2            1    1      0
## 3            1    1      1
## 4            1    1      0
## 5            1    0      1
## 6            1    0      0
## 7            1    0      1
## 8            1    0      0
## attr(,"assign")
## [1] 0 1 2
## attr(,"contrasts")
## attr(,"contrasts")$A
## [1] "contr.treatment"
##
## attr(,"contrasts")$B
## [1] "contr.treatment"
```

A model matrix for a model for a two-way factorial design with interaction term:

```
model.matrix(~ A * B, treats.df)
##    (Intercept) Ayes Bwhite Ayes:Bwhite
## 1            1    1      1           1
## 2            1    1      0           0
## 3            1    1      1           1
## 4            1    1      0           0
## 5            1    0      1           0
## 6            1    0      0           0
## 7            1    0      1           0
## 8            1    0      0           0
## attr(,"assign")
## [1] 0 1 2 3
## attr(,"contrasts")
## attr(,"contrasts")$A
## [1] "contr.treatment"
##
## attr(,"contrasts")$B
## [1] "contr.treatment"
```

4.10   Time series

Longitudinal data consist of repeated measurements, usually done over time, on the same experimental units. Longitudinal data, when replicated on several experimental units at each time point, are called repeated measurements, while when not replicated, they are called time series. Base R provides special support for the analysis of time series data, while repeated measurements can be analyzed with nested linear models, mixed-effects models, and additive models.

Time series data are data collected in such a way that there is only one observation, possibly of multiple variables, available at each point in time. This brief section introduces only the most basic aspects of time-series analysis. In most cases time steps are of uniform duration and occur regularly, which simplifies data handling and storage. R not only provides methods for the analysis and manipulation of time-series, but also a specialized class for their storage, "ts". Regular time steps allow more compact storage—e.g., a ts object does not need to store time values for each observation but instead a combination of two of start time, step size and end time.

We start by creating a time series from a numeric vector. By now, you surely guessed that you need to use a constructor called ts() or a conversion constructor called as.ts() and that you can look up the arguments they accept by reading the corresponding help pages.

For example for a time series of monthly values we could use:

```
my.ts <- ts(1:10, start = 2019, deltat = 1/12)
class(my.ts)
## [1] "ts"

str(my.ts)
##  Time-Series [1:10] from 2019 to 2020: 1 2 3 4 5 6 7 8 9 10
```

We next use the data set austres with data on the number of Australian residents and included in R.

```
class(austres)
## [1] "ts"

is.ts(austres)
## [1] TRUE
```

Time series austres is dominated by the increasing trend.

```
plot(austres)
```

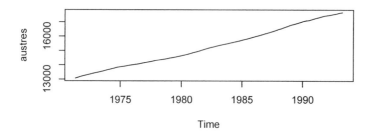

A different example, using data set `nottem` containing meteorological data for Nottingham, shows a clear cyclic component. The annual cycle of mean air temperatures (in degrees Fahrenheit) is clear when data are plotted.

```
data(nottem)
is.ts(nottem)
## [1] TRUE

plot(nottem)
```

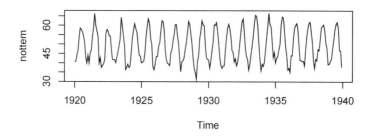

In the next two code chunks, two different approaches to time series decomposition are used. In the first one we use a moving average to capture the trend, while in the second approach we use Loess (a smooth curve fitted by local weighted regression) for the decomposition, a method for which the acronym STL (Seasonal and Trend decomposition using Loess) is used. Before decomposing the time-series we reexpress the temperatures in degrees Celsius.

```
nottem.celcius <- (nottem - 32) * 5/9
```

We set the seasonal window to 7 months, the minimum accepted.

```
nottem.stl <- stl(nottem.celcius, s.window = 7)
plot(nottem.stl)
```

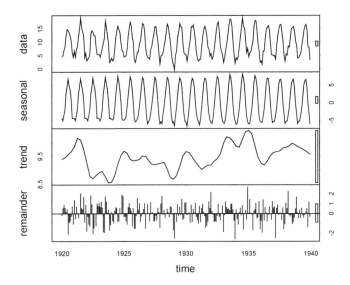

It is interesting to explore the class and structure of the object returned by `stl()`, as we may want to extract components. Run the statements below to find out, and then plot individual components from the time series decomposition.

```
class(nottem.stl)
str(nottem.stl)
```

4.11 Multivariate statistics

4.11.1 Multivariate analysis of variance

Multivariate methods take into account several response variables simultaneously, as part of a single analysis. In practice it is usual to use contributed packages for multivariate data analysis in R, except for simple cases. We will look first at *multivariate* ANOVA or MANOVA. In the same way as `aov()` is a wrapper that uses internally `lm()`, `manova()` is a wrapper that uses internally `aov()`.

Multivariate model formulas in base R require the use of column binding (`cbind()`) on the left-hand side (lhs) of the model formula. For the next examples we use the well-known `iris` data set, containing size measurements for flowers of two species of *Iris*.

```
data(iris)
mmf1 <- lm(cbind(Petal.Length, Petal.Width) ~ Species, data = iris)
anova(mmf1)
## Analysis of Variance Table
##
##                Df  Pillai approx F num Df den Df    Pr(>F)
## (Intercept)    1 0.98786   5939.2      2    146 < 2.2e-16 ***
## Species        2 1.04645     80.7      4    294 < 2.2e-16 ***
## Residuals    147
## ---
## Signif. codes:  0 '***' 0.001 '**' 0.01 '*' 0.05 '.' 0.1 ' ' 1

summary(mmf1)
## Response Petal.Length :
##
## Call:
## lm(formula = Petal.Length ~ Species, data = iris)
##
## Residuals:
##     Min      1Q Median     3Q    Max
## -1.260 -0.258  0.038  0.240  1.348
##
## Coefficients:
##                   Estimate Std. Error t value Pr(>|t|)
## (Intercept)        1.46200    0.06086   24.02   <2e-16 ***
## Speciesversicolor  2.79800    0.08607   32.51   <2e-16 ***
## Speciesvirginica   4.09000    0.08607   47.52   <2e-16 ***
## ---
## Signif. codes:  0 '***' 0.001 '**' 0.01 '*' 0.05 '.' 0.1 ' ' 1
##
## Residual standard error: 0.4303 on 147 degrees of freedom
## Multiple R-squared:  0.9414,Adjusted R-squared:  0.9406
## F-statistic:  1180 on 2 and 147 DF,  p-value: < 2.2e-16
##
##
## Response Petal.Width :
##
## Call:
## lm(formula = Petal.Width ~ Species, data = iris)
##
## Residuals:
##     Min      1Q Median     3Q    Max
## -0.626 -0.126 -0.026  0.154  0.474
##
## Coefficients:
##                   Estimate Std. Error t value Pr(>|t|)
## (Intercept)        0.24600    0.02894    8.50 1.96e-14 ***
## Speciesversicolor  1.08000    0.04093   26.39  < 2e-16 ***
## Speciesvirginica   1.78000    0.04093   43.49  < 2e-16 ***
## ---
## Signif. codes:  0 '***' 0.001 '**' 0.01 '*' 0.05 '.' 0.1 ' ' 1
##
## Residual standard error: 0.2047 on 147 degrees of freedom
## Multiple R-squared:  0.9289,Adjusted R-squared:  0.9279
## F-statistic:   960 on 2 and 147 DF,  p-value: < 2.2e-16


mmf2 <- manova(cbind(Petal.Length, Petal.Width) ~ Species, data = iris)
anova(mmf2)
```

```
## Analysis of Variance Table
##
##              Df  Pillai approx F num Df den Df    Pr(>F)
## (Intercept)   1 0.98786   5939.2      2    146 < 2.2e-16 ***
## Species       2 1.04645     80.7      4    294 < 2.2e-16 ***
## Residuals   147
## ---
## Signif. codes:  0 '***' 0.001 '**' 0.01 '*' 0.05 '.' 0.1 ' ' 1

summary(mmf2)
##           Df Pillai approx F num Df den Df    Pr(>F)
## Species    2 1.0465   80.661      4    294 < 2.2e-16 ***
## Residuals 147
## ---
## Signif. codes:  0 '***' 0.001 '**' 0.01 '*' 0.05 '.' 0.1 ' ' 1
```

> Modify the example above to use `aov()` instead of `manova()` and save the result to a variable named `mmf3`. Use `class()`, `attributes()`, `names()`, `str()` and extraction of members to explore objects `mmf1`, `mmf2` and `mmf3`. Are they different?

4.11.2   Principal components analysis

Principal components analysis (PCA) is used to simplify a data set by combining variables with similar and "mirror" behavior into principal components. At a later stage, we frequently try to interpret these components in relation to known and/or assumed independent variables. Base R's function `prcomp()` computes the principal components and accepts additional arguments for centering and scaling.

```
pc <- prcomp(iris[c("Sepal.Length", "Sepal.Width",
                    "Petal.Length", "Petal.Width")],
             center = TRUE,
             scale = TRUE)
```

By printing the returned object we can see the loadings of each variable in the principal components P1 to P4.

```
class(pc)
## [1] "prcomp"

pc
## Standard deviations (1, .., p=4):
## [1] 1.7083611 0.9560494 0.3830886 0.1439265
##
## Rotation (n x k) = (4 x 4):
##                     PC1         PC2         PC3        PC4
## Sepal.Length  0.5210659 -0.37741762  0.7195664  0.2612863
## Sepal.Width  -0.2693474 -0.92329566 -0.2443818 -0.1235096
## Petal.Length  0.5804131 -0.02449161 -0.1421264 -0.8014492
## Petal.Width   0.5648565 -0.06694199 -0.6342727  0.5235971
```

In the summary, the rows "Proportion of Variance" and "Cumulative Proportion" are most informative of the contribution of each principal component (PC) to explaining the variation among observations.

```
summary(pc)
## Importance of components:
##                           PC1    PC2     PC3     PC4
## Standard deviation     1.7084 0.9560 0.38309 0.14393
## Proportion of Variance 0.7296 0.2285 0.03669 0.00518
## Cumulative Proportion  0.7296 0.9581 0.99482 1.00000
```

Method `biplot()` produces a plot with one principal component (PC) on each axis, plus arrows for the loadings.

```
biplot(pc)
```

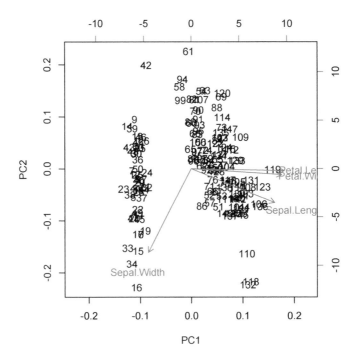

Method `plot()` generates a bar plot of variances corresponding to the different components.

```
plot(pc)
```

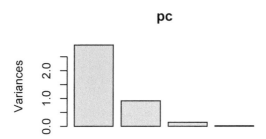

Visually more elaborate plots of the principal components and their loadings can be obtained using packages 'ggplot' described in chapter 7 starting on page 203. Package 'ggfortify' and package 'ggbiplot' extend 'ggplot' so as to make it easy to plot principal components and their loadings.

> For growth and morphological data, a log-transformation can be suitable given that variance is frequently proportional to the magnitude of the values measured. We leave as an exercise to repeat the above analysis using transformed values for the dimensions of petals and sepals. How much does the use of transformations change the outcome of the analysis?

> As for other fitted models, the object returned by function `prcomp()` is a list with multiple components.
>
> ```
> str(pc, max.level = 1)
> ```

4.11.3   Multidimensional scaling

The aim of multidimensional scaling (MDS) is to visualize in 2D space the similarity between pairs of observations. The values for the observed variable(s) are used to compute a measure of distance among pairs of observations. The nature of the data will influence what distance metric is most informative. For MDS we start with a matrix of distances among observations. We will use, for the example, distances in kilometers between geographic locations in Europe from data set `eurodist`.

```
loc <- cmdscale(eurodist)
```

We can see that the returned object `loc` is a `matrix`, with names for one of the dimensions.

```
class(loc)
## [1] "matrix" "array"

dim(loc)
## [1] 21  2

dimnames(loc)
## [[1]]
##  [1] "Athens"          "Barcelona"       "Brussels"         "Calais"
##  [5] "Cherbourg"       "Cologne"         "Copenhagen"       "Geneva"
##  [9] "Gibraltar"       "Hamburg"         "Hook of Holland"  "Lisbon"
## [13] "Lyons"           "Madrid"          "Marseilles"       "Milan"
## [17] "Munich"          "Paris"           "Rome"             "Stockholm"
## [21] "Vienna"
##
## [[2]]
## NULL

head(loc)
##                  [,1]       [,2]
## Athens     2290.27468 1798.8029
## Barcelona  -825.38279  546.8115
## Brussels     59.18334 -367.0814
## Calais      -82.84597 -429.9147
## Cherbourg -352.49943 -290.9084
## Cologne    293.68963 -405.3119
```

To make the code easier to read, two vectors are first extracted from the matrix and named x and y. We force aspect to equality so that distances on both axes are comparable.

```
x <- loc[, 1]
y <- -loc[, 2] # change sign so North is at the top
plot(x, y, type = "n", asp = 1,
     main = "cmdscale(eurodist)")
text(x, y, rownames(loc), cex = 0.6)
```

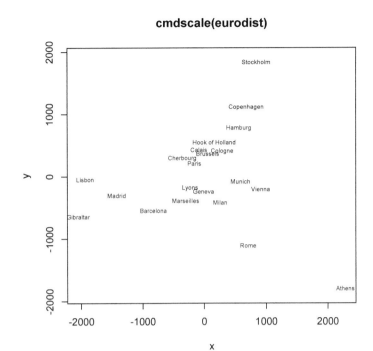

Find data on the mean annual temperature, mean annual rainfall and mean number of sunny days at each of the locations in the eurodist data set. Next, compute suitable distance metrics, for example, using function dist. Finally, use MDS to visualize how similar the locations are with respect to each of the three variables. Devise a measure of distance that takes into account the three climate variables and use MDS to find how distant the different locations are.

4.11.4   Cluster analysis

In cluster analysis, the aim is to group observations into discrete groups with maximal internal homogeneity and maximum group-to-group differences. In the next example we use function hclust() from the base-R package 'stats'. We use, as above, the eurodist data which directly provides distances. In other cases a matrix of distances between pairs of observations needs to be first calculated with function dist which supports several methods.

```
hc <- hclust(eurodist)
print(hc)
##
## Call:
## hclust(d = eurodist)
##
## Cluster method   : complete
```

```
## Number of objects: 21
```

```
plot(hc)
```

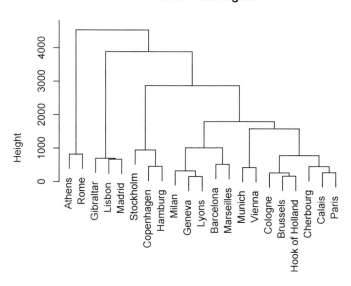

Cluster Dendrogram

eurodist
hclust (*, "complete")

We can use `cutree()` to limit the number of clusters by directly passing as an argument the desired number of clusters or the height at which to cut the tree.

```
cutree(hc, k = 5)
##              Athens       Barcelona        Brussels          Calais       Cherbourg
##                   1               2               3               3               3
##             Cologne      Copenhagen          Geneva       Gibraltar         Hamburg
##                   3               4               2               5               4
## Hook of Holland          Lisbon           Lyons          Madrid      Marseilles
##                   3               5               2               5               2
##               Milan          Munich           Paris            Rome       Stockholm
##                   2               3               3               1               4
##              Vienna
##                   3
```

The object returned by `hclust()` contains details of the result of the clustering, which allows further manipulation and plotting.

```
str(hc)
## List of 7
##  $ merge  : int [1:20, 1:2] -8 -3 -6 -4 -16 -17 -5 -7 -2 -12 ...
##  $ height : num [1:20] 158 172 269 280 328 428 460 460 521 668 ...
##  $ order  : int [1:21] 1 19 9 12 14 20 7 10 16 8 ...
##  $ labels : chr [1:21] "Athens" "Barcelona" "Brussels" "Calais" ...
##  $ method : chr "complete"
##  $ call   : language hclust(d = eurodist)
```

```
##   $ dist.method: NULL
##   - attr(*, "class")= chr "hclust"
```

4.12 Further reading

Two recent text books on statistics, following a modern approach, and using R for examples, are *OpenIntro Statistics* (Diez et al. 2019) and *Modern Statistics for Modern Biology* (Holmes and Huber 2019). Three examples of books introducing statistical computations in R are *Introductory Statistics with R* (Dalgaard 2008), *A Handbook of Statistical Analyses Using R* (B. S. Everitt and Hothorn 2009) and *A Beginner's Guide to R* (Zuur et al. 2009). More advanced books are available with detailed descriptions various types of analyses in R, including thorough descriptions of the methods briefly presented in this chapter. Good examples of books with broad scope are *The R Book* (Crawley 2012) and the classic reference *Modern Applied Statistics with S* (Venables and Ripley 2002). More specific books are also available from which a few suggestions for further reading are *An Introduction to Applied Multivariate Analysis with R* (B. Everitt and Hothorn 2011), *Linear Models with R* (Faraway 2004), *Extending the linear model with R: generalized linear, mixed effects and nonparametric regression models* (Faraway 2006), *Mixed-Effects Models in S and S-Plus* (Pinheiro and Bates 2000) and *Generalized Additive Models* (Wood 2017).

5

The R language: Adding new "words"

Computer Science is a science of abstraction—creating the right model for a problem and devising the appropriate mechanizable techniques to solve it.

Alfred V. Aho and Jeffrey D. Ullman
Foundations of Computer Science, 1992

5.1 Aims of this chapter

In earlier chapters we have only used base R features. In this chapter you will learn how to expand the range of features available. In the first part of the chapter we will focus on using existing packages and how they expand the functionality of R. In the second part you will learn how to define new functions, operators and classes. We will not consider the important, but more advanced question of packaging functions and classes into new R packages.

5.2 Packages

5.2.1 Sharing of R-language extensions

The most elegant way of adding new features or capabilities to R is through packages. This is without doubt the best mechanism when these extensions to R need to be shared. However, in most situations it is also the best mechanism for managing code that will be reused even by a single person over time. R packages have strict rules about their contents, file structure, and documentation, which makes it possible among other things for the package documentation to be merged into R's help system when a package is loaded. With a few exceptions, packages can be written so that they will work on any computer where R runs.

Packages can be shared as source or binary package files, sent for example through e-mail. However, for sharing packages widely, it is best to submit them to a repository. The largest public repository of R packages is called CRAN, an acronym for Comprehensive R Archive Network. Packages available through CRAN are guaranteed to work, in the sense of not failing any tests built into the package and not crashing or aborting prematurely. They are tested daily, as they may depend on other packages whose code will change when updated. In January 2017, the number of packages available through CRAN passed the 10,000 mark.

A key repository for bioinformatics with R is Bioconductor, containing packages that pass strict quality tests. Recently, ROpenScience has established guidelines and a system for code peer review for packages. These peer-reviewed packages are available through CRAN or other repositories and listed at the ROpenScience website. In some cases you may need or want to install less stable code from Git repositories such as versions still under development not yet submitted to CRAN. Using the package 'devtools' we can install packages directly from GitHub, Bitbucket and other code repositories based on Git. Installations from code repositories are always installations from sources (see below). It is of course also possible to install packages from local files (e.g., after a manual download).

One good way of learning how the extensions provided by a package work, is by experimenting with them. When using a function we are not yet familiar with, looking at its help to check all its features will expand your "toolbox." How much documentation is included with packages varies, while documentation of exported objects is enforced, many packages include, in addition, comprehensive user guides or articles as *vignettes*. It is not unusual to decide which package to use from a set of alternatives based on the quality of available documentation. In the case of packages adding extensive new functionality, they may be documented in depth in a book. Well-known examples are *Mixed-Effects Models in S and S-Plus* (Pinheiro and Bates 2000), *Lattice: Multivariate Data Visualization with R* (Sarkar 2008) and *ggplot2: Elegant Graphics for Data Analysis* (Wickham and Sievert 2016).

5.2.2   How packages work

The development of packages is beyond the scope of the current book, and thoroughly explained in the book *R Packages* (Wickham 2015). However, it is still worthwhile mentioning a few things about the development of R packages. Using RStudio it is relatively easy to develop your own packages. Packages can be of very different sizes and complexity. Packages use a relatively rigid structure of folders for storing the different types of files, including documentation compatible with R's built-in help system. This allows documentation for contributed packages to be seamlessly linked to R's help system when packages are loaded. In addition to R code, packages can call functions and routines written in C, C++, FORTRAN, Java, Python, etc., but some kind of "glue" is needed, as function call conventions and *name mangling* depend on the programming language, and in many cases also on the compiler used. For C++, the 'Rcpp' R package makes the "gluing" relatively easy (Eddelbuettel 2013). In the case of Python, R package 'reticulate' makes calling of Python methods and exchange of data easy, and it is well supported by RStudio. In the case of Java we can use package 'RJava' instead. For C and FORTRAN, R provides the functionality needed, but the interface needs some ad hoc coding in most cases.

Only objects exported by a package that has been attached are visible outside its own namespace. Loading and attaching a package with `library()` makes the exported objects available. Attaching a package adds the objects exported by the package to the search path so that they can be accessed without prepending the name of the namespace. Most packages do not export all the functions and objects defined in their code; some are kept internal, in most cases because they may change or be removed in future versions. Package namespaces can be detached and also unloaded with function `detach()` using a slightly different notation for the argument from that which we described for data frames in section 2.14.1 on page 71.

5.2.3 Download, installation and use

In R speak, "library" is the location where packages are installed. Packages are sets of functions, and data, specific for some particular purpose, that can be loaded into an R session to make them available so that they can be used in the same way as built-in R functions and data. Function `library()` is used to load and attach packages that are already installed in the local R library. In contrast, function `install.packages()` is used to install packages. When using RStudio it is easiest to use RStudio menus (which call `install.packages()` and `update.packages()`) to install or update packages.

> Use `help` to look up the help pages for `install.packages()` and `library()`, and explain what the code in the next chunk does.

R packages can be installed either from sources, or from already built "binaries". Installing from sources, depending on the package, may require additional software to be available. Under MS-Windows, the needed shell, commands and compilers are not available as part of the operating system. Installing them is not difficult as they are available prepackaged in installers (you will need RTools, and MiKTeX). It is easier to install packages from binary `.zip` files under MS-Windows. Under Linux most tools will be available, or very easy to install, so it is usual to install packages from sources. For OS X (Apple Mac) the situation is somewhere in-between. If the tools are available, packages can be very easily installed from sources from within RStudio. However, binaries are for most packages also readily available.

5.2.4 Finding suitable packages

Due to the large number of contributed R packages it can sometimes be difficult to find a suitable package for a task at hand. It is good to first check if the necessary capability is already built into base R. Base R plus the recommended packages (installed when R is installed) cover a lot of ground. To analyze data using almost any of the more common statistical methods does not require the use of special packages. Sometimes, contributed packages duplicate or extend the functionality in base R with advantage. When one considers the use of novel or specialized types of data analysis, the use of contributed packages can be unavoidable. Even in such cases, it is not unusual to have alternatives to choose from within the available contributed packages. Sometimes groups or suites of packages are designed to work well together.

The CRAN repository has very broad scope and includes a section called "views." R views are web pages providing annotated lists of packages frequently used within a given field of research, engineering or specific applications. These views are edited and updated by different editors. They can be found at `https://cran.r-project.org/web/views/`.

The Bioconductor repository specializes in bioinformatics with R. It also has a section with "views" and within it, descriptions of different data analysis workflows. The workflows are especially good as they reveal which sets of packages work well together. These views can be found at `https://www.bioconductor.org/packages/release/BiocViews.html`.

Although ROpenSci does not keep a separate package repository for the peer-reviewed packages, they do keep an index of them at `https://ropensci.org/packages/`.

The CRAN repository keeps an archive of earlier versions of packages, on an individual package basis. METACRAN (`https://www.r-pkg.org/`) is an archive of repositories, that keeps a historical record as snapshots from CRAN. METACRAN uses a different search engine than CRAN itself, making it easier to search the whole repository.

5.3 Defining functions and operators

Abstraction can be defined as separating the fundamental properties from the accidental ones. Say obtaining the mean from a given vector of numbers is an actual operation. There can be many such operations on different numeric vectors, each one a specific case. When we describe an algorithm for computing the mean from any numeric vector we have created the abstraction of *mean*. In the same way, each time we separate operations from specific data we create a new abstraction. In this sense, functions are abstractions of operations or actions; they are like "verbs" describing actions separately from actors.

The main role of functions is that of providing an abstraction allowing us to avoid repeating blocks of code (groups of statements) applying the same operations on different data. The reasons to avoid repetition of similar blocks of code statements are that 1) if the algorithm or implementation needs to be revised—e.g., to fix a bug or error—it is best to make edits in a single place; 2) sooner or later pieces of repeated code can become different leading to inconsistencies and hard-to-track bugs; 3) abstraction and division of a problem into smaller chunks, greatly helps with keeping the code understandable to humans; 4) textual repetition makes the script file longer, and this makes debugging, commenting, etc., more tedious, and error prone.

How do we, in practice, avoid repeating bits of code? We write a function containing the statements that we would need to repeat, and later we *call* ("use") the function in their place. We have been calling R functions or operators in almost every example in this book; what we will next tackle is how to define new functions of our own.

New functions and operators are defined using function `function()`, and saved like any other object in R by assignment to a variable name. In the example below, x and y are both formal parameters, or names used within the function for objects that will be supplied as *arguments* when the function is called. One can think of parameter names as placeholders for actual values to be supplied as arguments when calling the function.

```
my.prod <- function(x, y){x * y}
my.prod(4, 3)
## [1] 12
```

> ⚠ In base R, arguments to functions are passed by copy. This is something very important to remember. Whatever you do within a function to modify an argument, its value outside the function will remain (almost) always unchanged. (In other languages, arguments can also be passed by reference, meaning that assignments to a formal parameter within the body of the function are referenced to the argument and modify it. Such roundabout effects are frequently called side effects of a call. It is possible to imitate such behavior in R using some language trickery and consequently, some packages such as 'data.table' do define functions that use passing of arguments by reference.)
>
> ```
> my.change <- function(x){x <- NA}
> a <- 1
> my.change(a)
> a
> ## [1] 1
> ```
>
> In general, any result that needs to be made available outside the function must be returned by the function—or explicitly assigned to an object in the enclosing environment (i.e., using <<- or `assign()`) as a side effect.
>
> A function can only return a single object, so when multiple results are produced they need to be collected into a single object. In many cases, lists are used to collect all the values to be returned into one R object. For example, model fit functions like `lm()`, discussed in section 4.6 on page 127, return a complex list with heterogeneous named members.

> ℹ When function `return()` is called within a function, flow of execution within the function stops and the argument passed to `return()` is the value returned by the function call. In contrast, if function `return()` is not explicitly called, the value returned by the function call is that returned by the last statement *executed* within the body of the function.

```
print.x.1 <- function(x){print(x)}
print.x.1("test")
## [1] "test"

print.x.2 <- function(x){print(x); return(x)}
print.x.2("test")
## [1] "test"
## [1] "test"

print.x.3 <- function(x){return(x); print(x)}
print.x.3("test")
## [1] "test"

print.x.4 <- function(x){return(); print(x)}
print.x.4("test")
## NULL

print.x.5 <- function(x){x}
print.x.4("test")
## NULL
```

Test the behavior of functions `print.x.1()` and `print.x.5()`, as defined above, both at the command prompt, and in a script. The behavior of one of these functions will be different when the script is sourced than at the command prompt. Explain why.

Functions have their own scope. Any names created by normal assignment within the body of a function are visible only within the body of the function and disappear when the function returns from the call. In normal use, functions in R do not affect their environment through side effects. They receive input through arguments and return a value as the result of the call. This value can be either printed or assigned as we have seen when using functions earlier.

5.3.1 Ordinary functions

After the toy examples above, we will define a small but useful function: a function for calculating the standard error of the mean from a numeric vector. The standard error is given by $S_{\hat{x}} = \sqrt{S^2/n}$. We can translate this into the definition of an R function called SEM.

```
SEM <- function(x){sqrt(var(x) / length(x))}
```

We can test our function.

```
a <- c(1, 2, 3, -5)
a.na <- c(a, NA)
SEM(x = a)
## [1] 1.796988
```

```
SEM(a)
## [1] 1.796988

SEM(a.na)
## [1] NA
```

For example in SEM(a) we are calling function SEM() with a as an argument.

The function we defined above will always give the correct answer because NA values in the input will always result in an NA being returned. The problem is that unlike R's functions like var(), there is no option to omit NA values in the function we defined.

This could be implemented by adding a second parameter na.omit to the definition of our function and passing its argument to the call to var() within the body of SEM(). However, to avoid returning wrong values we need to make sure NA values are also removed before counting the number of observations with length().

A readable way of implementing this in code is to define the function as follows.

```
sem <- function(x, na.omit = FALSE) {
 if (na.omit) {
   x <- na.omit(x)
 }
 sqrt(var(x)/length(x))
}
```

```
sem(x = a)
## [1] 1.796988

sem(x = a.na)
## [1] NA

sem(x = a.na, na.omit = TRUE)
## [1] 1.796988
```

R does not provide a function for standard error, so the function above is generally useful. Its user interface is consistent with that of functionally similar existing functions. We have added a new word to the R vocabulary available to us.

In the definition of sem() we set a default argument for parameter na.omit which is used unless the user explicitly passes an argument to this parameter.

Define your own function to calculate the mean in a similar way as SEM() was defined above. Hint: function sum() could be of help.

Functions can have much more complex and larger compound statements as their body than those in the examples above. Within an expression, a function name followed by parentheses is interpreted as a call to the function. The bare name of a function instead gives access to its definition.

We first print (implicitly) the definition of our function from earlier in this section.

```
sem
## function(x, na.omit = FALSE) {
##   if (na.omit) {
##     x <- na.omit(x)
##   }
##   sqrt(var(x)/length(x))
## }
## <bytecode: 0x000000001c7dcd30>
```

Next we print the definition of R's linear model fitting function `lm()`. (Use of `lm()` is described in section 4.6 on page 127.)

```
lm
## function (formula, data, subset, weights, na.action, method = "qr",
##     model = TRUE, x = FALSE, y = FALSE, qr = TRUE, singular.ok = TRUE,
##     contrasts = NULL, offset, ...)
## {
##     ret.x <- x
##     ret.y <- y
##     cl <- match.call()
##     mf <- match.call(expand.dots = FALSE)
##     m <- match(c("formula", "data", "subset", "weights", "na.action",
##         "offset"), names(mf), 0L)
##     mf <- mf[c(1L, m)]
##     mf$drop.unused.levels <- TRUE
##     mf[[1L]] <- quote(stats::model.frame)
##     mf <- eval(mf, parent.frame())
##     if (method == "model.frame")
##         return(mf)
##     else if (method != "qr")
##         warning(gettextf("method = '%s' is not supported. Using 'qr'",
##             method), domain = NA)
##     mt <- attr(mf, "terms")
##     y <- model.response(mf, "numeric")
##     w <- as.vector(model.weights(mf))
##     if (!is.null(w) && !is.numeric(w))
##         stop("'weights' must be a numeric vector")
##     offset <- model.offset(mf)
##     mlm <- is.matrix(y)
##     ny <- if (mlm)
##         nrow(y)
##     else length(y)
##     if (!is.null(offset)) {
##         if (!mlm)
##             offset <- as.vector(offset)
##         if (NROW(offset) != ny)
##         stop(gettextf("number of offsets is %d, should equal %d (number of observations)",
##                 NROW(offset), ny), domain = NA)
##     }
##     if (is.empty.model(mt)) {
##         x <- NULL
##         z <- list(coefficients = if (mlm) matrix(NA_real_, 0,
##             ncol(y)) else numeric(), residuals = y, fitted.values = 0 *
##             y, weights = w, rank = 0L, df.residual = if (!is.null(w)) sum(w !=
##             0) else ny)
##         if (!is.null(offset)) {
##             z$fitted.values <- offset
##             z$residuals <- y - offset
##         }
```

```
##       }
##       else {
##           x <- model.matrix(mt, mf, contrasts)
##           z <- if (is.null(w))
##               lm.fit(x, y, offset = offset, singular.ok = singular.ok,
##                   ...)
##           else lm.wfit(x, y, w, offset = offset, singular.ok = singular.ok,
##               ...)
##       }
##       class(z) <- c(if (mlm) "mlm", "lm")
##       z$na.action <- attr(mf, "na.action")
##       z$offset <- offset
##       z$contrasts <- attr(x, "contrasts")
##       z$xlevels <- .getXlevels(mt, mf)
##       z$call <- cl
##       z$terms <- mt
##       if (model)
##           z$model <- mf
##       if (ret.x)
##           z$x <- x
##       if (ret.y)
##           z$y <- y
##       if (!qr)
##           z$qr <- NULL
##       z
## }
## <bytecode: 0x00000000151d2418>
## <environment: namespace:stats>
```

As can be seen at the end of the listing, this function written in the R language has been byte-compiled so that it executes faster. Functions that are part of the R language, but that are not coded using the R language, are called primitives and their full definition cannot be accessed through their name (c.f., sem() defined above).

```
list
## function (...)  .Primitive("list")
```

5.3.2   Operators

Operators are functions that use a different syntax for being called. If their name is enclosed in back ticks they can be called as ordinary functions. Binary operators like + have two formal parameters, and unary operators like unary – have only one formal parameter. The parameters of many binary R operators are named e1 and e2.

```
1 / 2
## [1] 0.5

`/`(1 , 2)
## [1] 0.5

`/`(e1 = 1 , e2 = 2)
## [1] 0.5
```

An important consequence of the possibility of calling operators using ordinary syntax is that operators can be used as arguments to *apply* functions in the same way as ordinary functions. When passing operator names as arguments to *apply* functions we only need to enclose them in back ticks (see section 3.4 on page 108).

The name by itself and enclosed in back ticks allows us to access the definition of an operator.

```
`/`
## function (e1, e2)  .Primitive("/")
```

Defining a new operator. We will define a binary operator (taking two arguments) that subtracts from the numbers in a vector the mean of another vector. First we need a suitable name, but we have less freedom as names of user-defined operators must be enclosed in percent signs. We will use `%-mean%` and as with any *special name*, we need to enclose it in quotation marks for the assignment.

```
"%-mean%" <- function(e1, e2) {
  e1 - mean(e2)
}
```

We can then use our new operator in a example.

```
10:15 %-mean% 1:20
## [1] -0.5  0.5  1.5  2.5  3.5  4.5
```

To print the definition, we enclose the name of our new operator in back ticks—i.e., we *back quote* the special name.

```
`%-mean%`
## function(e1, e2) {
##    e1 - mean(e2)
## }
```

5.4   Objects, classes, and methods

New classes are normally defined within packages rather than in user scripts. To be really useful implementing a new class involves not only defining a class but also a set of specialized functions or *methods* that implement operations on objects belonging to the new class. Nevertheless, an understanding of how classes work is important even if only very occasionally a user will define a new method for an existing class within a script.

Classes are abstractions, but abstractions describing the shared properties of "types" or groups of similar objects. In this sense, classes are abstractions of "actors," they are like "nouns" in natural language. What we obtain with classes is the possibility of defining multiple versions of functions (or *methods*) sharing the same name but tailored to operate on objects belonging to different classes. We have already been using methods with multiple *specializations* throughout the book, for example `plot()` and `summary()`.

We start with a quotation from *S Poetry* (Burns 1998, page 13).

> The idea of object-oriented programming is simple, but carries a lot of weight. Here's the whole thing: if you told a group of people "dress for work," then you would expect each to put on clothes appropriate for that individual's job. Likewise it is possible for S[R] objects to get dressed appropriately depending on what class of object they are.

We say that specific methods are *dispatched* based on the class of the argument passed. This, together with the loose type checks of R, allows writing code that functions as expected on different types of objects, e.g., character and numeric vectors.

R has good support for the object-oriented programming paradigm, but as a system that has evolved over the years, currently R supports multiple approaches. The still most popular approach is called S3, and a more recent and powerful approach, with slower performance, is called S4. The general idea is that a name like "plot" can be used as a generic name, and that the specific version of `plot()` called depends on the arguments of the call. Using computing terms we could say that the *generic* of `plot()` dispatches the original call to different specific versions of `plot()` based on the class of the arguments passed. S3 generic functions dispatch, by default, based only on the argument passed to a single parameter, the first one. S4 generic functions can dispatch the call based on the arguments passed to more than one parameter and the structure of the objects of a given class is known to the interpreter. In S3 functions, the specializations of a generic are recognized/identified only by their name. And the class of an object by a character string stored as an attribute to the object.

We first explore one of the methods already available in R. The definition of `mean` shows that it is the generic for a method.

```
mean
## function (x, ...)
## UseMethod("mean")
## <bytecode: 0x00000000138ddc60>
## <environment: namespace:base>
```

We can find out which specializations of method are available in the current search path using `methods()`.

```
methods(mean)
## [1] mean.Date     mean.default  mean.difftime mean.POSIXct  mean.POSIXlt
## see '?methods' for accessing help and source code
```

We can also use `methods()` to query all methods, including operators, defined for objects of a given class.

```
methods(class = "list")
## [1] all.equal       as.data.frame coerce         Ops           relist
## [6] type.convert    within
## see '?methods' for accessing help and source code
```

S3 class information is stored as a character vector in an attribute named "class". The most basic approach to creation of an object of a new S3 class, is to add the new class name to the class attribute of the object. As the implied class hierarchy is given by the order of the members of the character vector, the name of the new class must be added at the head of the vector. Even though this step can be done as shown here, in practice this step would normally take place within a *constructor* function and the new class, if defined within a package, would need to be registered. We show here this bare-bones example to demonstrate how S3 classes are implemented in R.

```
a <- 123
class(a)
## [1] "numeric"

class(a) <- c("myclass", class(a))
class(a)
## [1] "myclass" "numeric"
```

Now we create a print method specific to "myclass" objects. Internally we are using function sprintf() and for the format template to work we need to pass a numeric value as an argument—i.e., obviously sprintf() does not "know" how to handle objects of the class we have just created!

```
print.myclass <- function(x) {
    sprintf("[myclass] %.0f", as.numeric(x))
}
```

Once a specialized method exists for a class, it will be used for objects of this class.

```
print(a)
## [1] "[myclass] 123"

print(as.numeric(a))
## [1] 123
```

> 📖 The S3 class system is "lightweight" in that it adds very little additional computation load, but it is rather "fragile" in that most of the responsibility for consistency and correctness of the design—e.g., not messing up dispatch by redefining functions or loading a package exporting functions with the same name, etc., is not checked by the R interpreter.
>
> Defining a new S3 generic is also quite simple. A generic method and a default method need to be created.

```
my_print <- function (x, ...) {
   UseMethod("my_print", x)
 }

my_print.default <- function(x, ...) {
   print(class(x))
   print(x, ...)
}
```

```
my_print(123)
## [1] "numeric"
## [1] 123
```

```
my_print("abc")
## [1] "character"
## [1] "abc"
```

Up to now, `my_print()`, has no specialization. We now write one for data frames.

```
my_print.data.frame <- function(x, rows = 1:5, ...) {
   print(x[rows, ], ...)
   invisible(x)
}
```

We add the second statement so that the function invisibly returns the whole data frame, rather than the lines printed. We now do a quick test of the function.

```
my_print(cars)
##   speed dist
## 1     4    2
## 2     4   10
## 3     7    4
## 4     7   22
## 5     8   16
```

```
my_print(cars, 8:10)
##    speed dist
## 8     10   26
## 9     10   34
## 10    11   17
```

```
b <- my_print(cars)
##    speed dist
## 1     4    2
## 2     4   10
## 3     7    4
## 4     7   22
## 5     8   16

str(b)
## 'data.frame': 50 obs. of  2 variables:
##  $ speed: num  4 4 7 7 8 9 10 10 10 11 ...
##  $ dist : num  2 10 4 22 16 10 18 26 34 17 ...

nrow(b) == nrow(cars) # was the whole data frame returned?
## [1] TRUE
```

5.5   Scope of names

The visibility of names is determined by the *scoping rules* of a language. The clearest, but not the only situation when scoping rules matter, is when objects with the same name coexist. In such a situation one will be accessible by its unqualified name and the other hidden but possibly accessible by qualifying the name with its name space.

As the R language has few reserved words for which no redefinition is allowed, we should take care not to accidentally reuse names that are part of language. For example pi is a constant defined in R with the value of the mathematical constant π. If we use the same name for one of our variables, the original definition becomes hidden.

```
pi
## [1] 3.141593

pi <- "apple pie"
pi
## [1] "apple pie"

rm(pi)
pi
## [1] 3.141593

exists("pi")
## [1] TRUE
```

In the example above, the two variables are not defined in the same scope. In the example below we assign a new value to a variable we have earlier created within the same scope, and consequently the second assignment overwrites, rather than hides, the existing definition.

```
my.pie <- "raspberry pie"
my.pie
## [1] "raspberry pie"

my.pie <- "apple pie"
my.pie
## [1] "apple pie"

rm(my.pie)
exists("my.pie")
## [1] FALSE
```

An additional important thing to remember is that R packages define all objects within a *namespace* with the same name as the package itself. This means that when we reuse a name defined in a package, its definition in the package does not get overwritten, but instead, only hidden and still accessible using the name *qualified* by prepending the name of the package followed by two colons.

If two packages define objects with the same name, then which one is visible depends on the order in which the packages were attached. To avoid confusion in such cases, in scripts is best to use the qualified names for calling both definitions.

5.6 Further reading

An in-depth discussion of object-oriented programming in R is outside the scope of this book. For the non-programmer user, a basic understanding of R classes can be useful, even if he or she does not intend to create new classes. This basic knowledge is what we covered in this chapter. Several books describe in detail the different class systems available and how to use them in R packages. For an in-depth treatment of the subject please consult the books *Advanced R* (Wickham 2019) and *Extending R* (Chambers 2016).

The development of packages is thoroughly described in the book *R Packages* (Wickham 2015) and an in-depth description of R from the programming perspective is given in the book *Advanced R* (Wickham 2019). The book *Extending R* (Chambers 2016) covers both subjects.

6

New grammars of data

Essentially everything in S[R], for instance, a call to a function, is an S[R] object. One viewpoint is that S[R] has self-knowledge. This self-awareness makes a lot of things possible in S[R] that are not in other languages.

Patrick J. Burns
S Poetry, 1998

6.1 Aims of this chapter

Base R and the recommended extension packages (installed by default) include many functions for manipulating data. The R distribution supplies a complete set of functions and operators that allow all the usual data manipulation operations. These functions have stable and well-described behavior, so they should be preferred unless some of their limitations justify the use of alternatives defined in contributed packages. In the present chapter we aim at describing the new syntaxes introduced by the most popular of these contributed R extension packages aiming at changing (usually improving one aspect at the expense of another) in various ways how we can manipulate data in R. These independently developed packages extend the R language not only by adding new "words" to it but by supporting new ways of meaningfully connecting "words"—i.e., providing new "grammars" for data manipulation.

6.2 Introduction

By reading previous chapters, you have already become familiar with base R classes, methods, functions and operators for storing and manipulating data. Most of these had been originally designed to perform optimally on rather small data sets (see Matloff 2011). The R implementation has been improved over the years significantly in performance, and random-access memory in computers has become cheaper, making constraints imposed by the original design of R less limiting, but on the other hand, the size of data sets has also increased. Some contributed packages have aimed at improving performance by relying on different compromises between usability, speed and reliability than used for base R.

Package 'data.table' is the best example of an alternative implementation of data storage that maximizes the speed of processing for large data sets using a new semantics and requiring a new syntax. We could say that package 'data.table' is based on a "grammar of data" that is different from that in the R language. The compromise in this case has been the use of a less intuitive syntax, and by defaulting to call by reference of arguments instead of by copy, increasing the "responsibility" of the author of code defining new functions.

When a computation includes a chain of sequential operations, if using base R, we can either store at each step in the computation the returned value in a variable, or nest multiple function calls. The first approach is verbose, but allows readable scripts, especially if variable names are wisely chosen. The second approach becomes very difficult too read as soon as there is more than one nesting level. Attempts to find an alternative syntax have borrowed the concept of data *pipes* from Unix shells (Kernigham and Plauger 1981). Interestingly, that it has been possible to write packages that define the operators needed to "add" this new syntax to R is a testimony to its flexibility and extensibility. Two packages, 'magrittr' and 'wrapr', define operators for pipe-based syntax.

A different aspect of the R syntax is extraction of members from lists and data frames by name. Base R provides two different operators for this, $ and [[]], with different syntax. These two operators also differ in how *incomplete names* are handled. Package 'tibble' alters this syntax for an alternative to base R's data frames. Once again, a new syntax allows new functionality at the expense of partial incompatibility with base R syntax. Objects of class "tb" were also an attempt to improve performance compared to objects of class "data.frame". R performance has improved in recent releases and currently, even though performance is not the same, depending on the operations and data, either R's data frames or tibbles perform better.

Base R function subset() has an unusual syntax, as it evaluates the expression passed as the second argument within the namespace of the data frame passed as its first argument (see 2.14.1 on page 71). This saves typing at the expense of increasing the risk of bugs, as by reading the call to subset, it is not obvious which names are resolved in the environment of subset() and which ones within its first argument—i.e., as column names in the data frame. In addition, changes elsewhere in a script can change how a call to subset is interpreted. In reality, subset is a wrapper function built on top of the extraction operator []. It is a convenience function, mostly intended to be used at the console, rather than in scripts or package code. To extract rows from a data frame it is always best to use the [,] operator.

Package 'dplyr' provides convenience functions that work in a similar way as base R subset(), although in latest versions more safely. This package has suffered quite drastic changes during its development with respect to how to handle the dilemma caused by "guessing" of the environment where names should be looked up. There is no easy answer; a simplified syntax leads to ambiguity, and a fully specified syntax is verbose. Recent versions of the package introduced a terse syntax to achieve a concise way of specifying where to look up names. My opinion is that for code that needs to be highly reliable and produce reproducible results in the future, we should for the time being prefer base R. For code that is to be used once, or for which reproducibility can depend on the use of a specific (old or soon to be old) version of 'dplyr', or which is not a burden to update, the conciseness and power of the new syntax will be an advantage.

In this chapter you will become familiar with alternative "grammars of data" as implemented in some of the packages that enable new approaches to manipulating data in R. As in previous chapters I will focus more on the available tools and how to use them than on their role in the analysis of data. The books *R for Data Science* (Wickham and Grolemund 2017) and *R Programming for Data Science* (Peng 2016) partly cover the same subjects from the perspective of data analysis.

6.3 Packages used in this chapter

```
install.packages(learnrbook::pkgs_ch_data)
```

To run the examples included in this chapter, you need first to load some packages from the library (see section 5.2 on page 163 for details on the use of packages).

```
library(learnrbook)
library(tibble)
library(magrittr)
library(wrapr)
library(stringr)
library(dplyr)
library(tidyr)
library(lubridate)
```

6.4 Replacements for `data.frame`

6.4.1 'data.table'

The function call semantics of the R language is that arguments are passed to functions by copy. If the arguments are modified within the code of a function, these changes are local to the function. If implemented naively, this semantic would impose a huge toll on performance, however, R in most situations only makes a copy if and when the value changes. Consequently, for modern versions of R which are very good at avoiding unnecessary copying of objects, the normal R semantics has only a moderate negative impact on performance. However, this impact can still be a problem as modification is detected at the object level, and consequently R may make copies of a whole data frame when only values in a single column or even just an attribute have changed.

Functions and methods from package 'data.table' use arguments by reference, avoiding making any copies. However, any assignments within these functions and methods modify the variable passed as an argument. This simplifies the needed tests for delayed copying and also by avoiding the need to make a copy of arguments, achieves the best possible performance. This is a specialized package but extremely useful when dealing with very large data sets. Writing user code, such as scripts, with 'data.table' requires a good understanding of the pass-by-reference semantics. Obviously, package 'data.table' makes no attempt at backwards compatibility with base-R data.frame.

6.4.2 'tibble'

The authors of package 'tibble' describe their tbl class as backwards compatible with data.frame and make it a derived class. This backwards compatibility is only partial so in some situations data frames and tibbles are not equivalent.

The class and methods that package 'tibble' defines lift some of the restrictions imposed by the design of base R data frames at the cost of creating some incompatibilities due to changed (improved) syntax for member extraction and by adding support for "columns" of class list and removing support for columns of class matrix. Handling of attributes is also different, with no row names added by default. There are also differences in default behavior of both constructors and methods. Although, objects of class tbl can be passed as arguments to functions that expect data frames as input, these functions are not guaranteed to work correctly as a result of the differences in syntax.

⚠ It is easy to write code that will work correctly both with data frames and tibbles. However, code that is syntactically correct according to the R language may fail if a tibble is used in place of a data frame.

🖳 The print() method for tibbles differs from that for data frames in that it outputs a header with the text "A tibble:" followed by the dimensions (number of rows × number of columns), adds under each column name an abbreviation of its class and instead of printing all rows and columns, a limited number of them are displayed. In addition, individual values are formatted differently even adding color highlighting for negative numbers.

```
tibble(A = LETTERS[1:5], B = -2:2, C = seq(from = 1, to = 0, length.out = 5))
## # A tibble: 5 x 3
##     A       B     C
##   <chr> <int> <dbl>
## 1 A      -2   1
## 2 B      -1   0.75
## 3 C       0   0.5
## 4 D       1   0.25
## 5 E       2   0
```

The default number of rows printed can be set with options, that we set here to only three rows for most of this chapter.

```
options(tibble.print_max = 3, tibble.print_min = 3)
```

ⓘ In their first incarnation, the name for `tibble` was `data_frame` (with a dash instead of a dot). The old name is still recognized, but its use should be avoided and `tibble()` used instead. One should be aware that although the constructor `tibble()` and conversion function `as_tibble()`, as well as the test `is_tibble()` use the name `tibble`, the class attribute is named `tbl`.

```
my.tb <- tibble(numbers = 1:3)
is_tibble(my.tb)
## [1] TRUE

inherits(my.tb, "tibble")
## [1] FALSE

class(my.tb)
## [1] "tbl_df"      "tbl"            "data.frame"
```

Furthermore, by necessity, to support tibbles based on different underlying data sources, a further derived class is needed. In our example, as our tibble has an underlying `data.frame` class, the most derived class of `my.tb` is `tbl_df`.

We start with the constructor and conversion methods. For this we will define our own diagnosis function (*apply* functions are described in section 3.4 on page 108).

```
show_classes <- function(x) {
  cat(
    paste(paste(class(x)[1],
      "containing:"),
    paste(names(x),
          sapply(x, class), collapse = ", ", sep = ": "),
    sep = "\n")
  )
}
```

In the next two chunks we can see some of the differences. The `tibble()`

constructor does not by default convert character data into factors, while the `data.frame()` constructor does.

```
my.df <- data.frame(codes = c("A", "B", "C"), numbers = 1:3, integers = 1L:3L)
is.data.frame(my.df)
## [1] TRUE

is_tibble(my.df)
## [1] FALSE

show_classes(my.df)
## data.frame containing:
## codes: character, numbers: integer, integers: integer
```

Tibbles are data frames—or more formally class `tibble` is derived from class `data.frame`. However, data frames are not tibbles.

```
my.tb <- tibble(codes = c("A", "B", "C"), numbers = 1:3, integers = 1L:3L)
is.data.frame(my.tb)
## [1] TRUE

is_tibble(my.tb)
## [1] TRUE

show_classes(my.tb)
## tbl_df containing:
## codes: character, numbers: integer, integers: integer
```

The `print()` method for tibbles, overrides the one defined for data frames.

```
print(my.df)
##   codes numbers integers
## 1     A       1        1
## 2     B       2        2
## 3     C       3        3

print(my.tb)
## # A tibble: 3 x 3
##   codes numbers integers
##   <chr>   <int>    <int>
## 1 A           1        1
## 2 B           2        2
## 3 C           3        3
```

> 🔋 Tibbles and data frames differ in how they are printed when they have many rows or columns. 1) Construct a data frame and an equivalent tibble with at least 50 rows and then test how the output looks when they are printed. 2) Construct a data frame and an equivalent tibble with more columns than will fit in the width of the Rconsole and then test how the output looks when they are printed.

Data frames can be converted into tibbles with `as_tibble()`.

```
my_conv.tb <- as_tibble(my.df)
is.data.frame(my_conv.tb)
## [1] TRUE

is_tibble(my_conv.tb)
## [1] TRUE

show_classes(my_conv.tb)
## tbl_df containing:
## codes: character, numbers: integer, integers: integer


my_conv.df <- as.data.frame(my.tb)
is.data.frame(my_conv.df)
## [1] TRUE

is_tibble(my_conv.df)
## [1] FALSE

show_classes(my_conv.df)
## data.frame containing:
## codes: character, numbers: integer, integers: integer
```

Look carefully at the result of the conversions. Why do we now have a data frame with A as character and a tibble with A as a factor?

Not all conversion functions work consistently when converting from a derived class into its parent. The reason for this is disagreement between authors on what the *correct* behavior is based on logic and theory. You are not likely to be hit by this problem frequently, but it can be difficult to diagnose.

We have already seen that calling as.data.frame() on a tibble strips the derived class attributes, returning a data frame. We will look at the whole character vector stored in the "class" attribute to demonstrate the difference. We also test the two objects for equality, in two different ways. Using the operator == tests for equivalent objects. Objects that contain the same data. Using identical() tests that objects are exactly the same, including attributes such as "class", which we retrieve using class().

```
class(my.tb)
## [1] "tbl_df"       "tbl"           "data.frame"

class(my_conv.df)
## [1] "data.frame"

my.tb == my_conv.df
##       codes numbers integers
## [1,]  TRUE   TRUE    TRUE
## [2,]  TRUE   TRUE    TRUE
## [3,]  TRUE   TRUE    TRUE

identical(my.tb, my_conv.df)
## [1] FALSE
```

Now we derive from a tibble, and then attempt a conversion back into a tibble.

```
my.xtb <- my.tb
class(my.xtb) <- c("xtb", class(my.xtb))
class(my.xtb)
## [1] "xtb"           "tbl_df"       "tbl"           "data.frame"

my_conv_x.tb <- as_tibble(my.xtb)
class(my_conv_x.tb)
## [1] "tbl_df"       "tbl"           "data.frame"

my.xtb == my_conv_x.tb
##       codes numbers integers
## [1,]  TRUE   TRUE    TRUE
## [2,]  TRUE   TRUE    TRUE
## [3,]  TRUE   TRUE    TRUE

identical(my.xtb, my_conv_x.tb)
## [1] FALSE
```

The two viewpoints on conversion functions are as follows. 1) The conversion function should return an object of its corresponding class, even if the argument is an object of a derived class, stripping the derived class. 2) If the object is of the class to be converted to, including objects of derived classes, then it should remain untouched. Base R follows, as far as I have been able to work out, approach 1). Packages in the 'tidyverse' follow approach 2). If in doubt about the behavior of some function, then you will need to do a test similar to the one used in this box.

There are additional important differences between the constructors `tibble()` and `data.frame()`. One of them is that in a call to `tibble()`, member variables ("columns") being defined can be used in the definition of subsequent member variables.

```
tibble(a = 1:5, b = 5:1, c = a + b, d = letters[a + 1])
## # A tibble: 5 x 4
##       a      b      c d
```

```
##      <int> <int> <int> <chr>
## 1       1     5     6 b
## 2       2     4     6 c
## 3       3     3     6 d
## # ... with 2 more rows
```

> 📖 What is the behavior if you replace `tibble()` by `data.frame()` in the statement above?

While data frame columns can be factors, vectors or matrices (with the same number of rows as the data frame), columns of tibbles can be factors, vectors or lists (with the same number of members as rows the tibble has).

```
tibble(a = 1:5, b = 5:1, c = list("a", 2, 3, 4, 5))
## # A tibble: 5 x 3
##       a     b c
##   <int> <int> <list>
## 1     1     5 <chr [1]>
## 2     2     4 <dbl [1]>
## 3     3     3 <dbl [1]>
## # ... with 2 more rows
```

Which even allows a list of lists as a variable, or a list of vectors.

```
tibble(a = 1:5, b = 5:1, c = list("a", 1:2, 0:3, letters[1:3], letters[3:1]))
## # A tibble: 5 x 3
##       a     b c
##   <int> <int> <list>
## 1     1     5 <chr [1]>
## 2     2     4 <int [2]>
## 3     3     3 <int [4]>
## # ... with 2 more rows
```

6.5 Data pipes

The first obvious difference between scripts using some of the new grammars is the frequent use of *pipes*. This is, however, mostly a question of preferences, as pipes can be used equally well with base R functions. Pipes have been at the core of shell scripting in Unix since early stages of its design (Kernigham and Plauger 1981). Within an OS, pipes are chains of small programs or "tools" that carry out a single well-defined task (e.g., `ed`, `gsub`, `grep`, `more`, etc.). Data such as text is described as flowing from a source into a sink through a series of steps at which a specific transformation takes place. In Unix, sinks and sources are files, but files as an abstraction include all devices and connections for input or output, including physical ones as terminals and printers. The connection between steps in the pipe is usually implemented by means of temporary files.

```
stdin | grep("abc") | more
```

How can *pipes* exist within a single R script? When chaining functions into a pipe, data is passed between them through temporary R objects stored in memory, which are created and destroyed automatically. Conceptually there is little difference between Unix shell pipes and pipes in R scripts, but the implementations are different.

What do pipes achieve in R scripts? They relieve the user from the responsibility of creating and deleting the temporary objects and of enforcing the sequential execution of the different steps. Pipes usually improve readability of scripts by allowing more concise code.

Currently, two main implementations of pipes are available as R extensions, in packages 'magrittr' and 'wrapr'.

6.5.1   'magrittr'

One set of operators needed to build pipes of R functions is implemented in package 'magrittr'. This implementation is used in the 'tidyverse' and the pipe operator re-exported by package 'dplyr'.

We start with a toy example first written using separate steps and normal R syntax

```
data.in <- 1:10
data.tmp <- sqrt(data.in)
data.out <- sum(data.tmp)
rm(data.tmp) # clean up!
```

next using nested function calls still using normal R syntax

```
data.out <- sum(sqrt(data.in))
```

written as a pipe using the chaining operator from package 'magrittr'.

```
data.in %>% sqrt() %>% sum() -> data.out
```

    The %>% from package 'magrittr' takes two operands. The value returned by the *lhs* (left-hand side) operand, which can be any R expression, is passed as first argument to the *rhs* operand, which must be a function accepting at least one argument. Consequently, in this implementation, the function in the *rhs* must have a suitable signature for the pipe to work implicitly as usually used. However, it is possible to pass piped arguments to a function by name or to other parameters than the first one using a dot (.) as placeholder.

    Some base R functions like subset() have a signature that is suitable for use in 'magrittr' pipes using implicit passing of the piped value to the first argument, while others such as assign() will not. In such cases we can use . as a placeholder and pass it as an argument, or, alternatively, define a wrapper function to change the order of the formal parameters in the function signature.

Package 'magrittr' provides additional pipe operators, such as "tee" (%T>%) to create a branch in the pipe, and %<>% to apply the pipe by reference. These operators are less frequently used than %>%.

6.5.2 'wrapr'

The %.>%, or "dot-pipe" operator from package 'wrapr', allows expressions both on the rhs and lhs, and enforces the use of the dot (.), as placeholder for the piped object.

Rewritten using the dot-pipe operator, the pipe in the previous chunk becomes

```
data.in %.>% sqrt(.) %.>% sum(.) -> data1.out
```

However, the same code can use the pipe operator from 'magrittr'.

```
data.in %>% sqrt(.) %>% sum(.) -> data2.out
all.equal(data1.out, data2.out)
## [1] TRUE
```

If needed or desired, named arguments are supported with the dot-pipe operator resulting in the expected behavior.

```
data.in %.>% assign(value = ., x = "data3.out")
all.equal(data.in, data3.out)
## [1] TRUE
```

In contrast, the pipe operator silently and unexpectedly fails to create the variable for the same example.

```
data.in %>% assign(value = ., x = "data4.out")
exists("data4.out")
## [1] FALSE
```

The dot-pipe operator allows us to use . in expressions as shown below, while %>% fails with an error (not shown).

```
data.in %.>% (2 + .^2) %.>% assign("data1.out", .)
```

> In conclusion, R syntax for expressions is preserved when using the dot-pipe operator, with the only caveat that because of the higher precedence of the %.>% operator, we need to "protect" bare expressions containing other operators by enclosing them in parentheses.

Under-the-hood, the implementations of %>% and %.>% are very different, with %.>% usually having better performance.

In the rest of the book we will exclusively use *dot pipes* in examples to ensure easier understanding as they avoid implicit ("invisible") passing of arguments and impose fewer restrictions on the syntax that can be used.

Although pipes can make scripts visually very different from the use of assignments of intermediate results to variables, from the point of view of data analysis what makes pipes most convenient to use are some of the new classes, functions, and methods defined in 'tidyr', 'dplyr', and other packages from the 'tidyverse'.

6.6   Reshaping with 'tidyr'

Data stored in table-like formats can be arranged in different ways. In base R most model fitting functions and the `plot()` method using (model) formulas and accepting data frames, expect data to be arranged in "long form" so that each row in a data frame corresponds to a single observation (or measurement) event on a subject. Each column corresponds to a different measured feature, time of measurement, or a factor describing a classification of subjects according to treatments or features of the experimental design (e.g., blocks). Covariates measured on the same subject at an earlier point in time may also be stored in a column. Data arranged in *long form* has been nicknamed as "tidy" and this is reflected in the name given to the 'tidyverse' suite of packages. Data in which columns correspond to measurement events is described as being in a *wide form*.

Although long-form data is and has been the most commonly used arrangement of data in R, manipulation of such data has not always been possible with concise R statements. The packages in the 'tidyverse' provide convenience functions to simplify coding of data manipulation, which in some cases, have, in addition, improved performance compared to base R—i.e., it is possible to code the same operations using only base R, but may require more and/or more verbose statements.

Real-world data is rather frequently stored in wide format or even ad hoc formats, so in many cases the first task in data analysis is to reshape the data. Package 'tidyr' provides functions for reshaping data from wide to long form and *vice versa* (replacing the older packages 'reshape' and 'reshape2').

We use in examples below the `iris` data set included in base R. Some operations on R `data.frame` objects with 'tidyverse' packages will return `data.frame` objects while others will return tibbles—i.e., "tb" objects. Consequently it is safer to first convert into tibbles the data frames we will work with.

```
iris.tb <- as_tibble(iris)
```

Function `gather()` converts data from wide form into long form (or "tidy"). We use `gather` to obtain a long-form tibble. By comparing `iris.tb` with `long_iris.tb` we can appreciate how `gather()` reshaped its input.

```
head(iris.tb, 2)
## # A tibble: 2 x 5
##   Sepal.Length Sepal.Width Petal.Length Petal.Width Species
##          <dbl>       <dbl>        <dbl>       <dbl> <fct>
## 1          5.1         3.5          1.4         0.2 setosa
## 2          4.9         3            1.4         0.2 setosa

iris.tb %.>%
```

```
  gather(., key = part, value = dimension, -Species) -> long_iris.tb
long_iris.tb
## # A tibble: 600 x 3
##    Species part           dimension
##    <fct>   <chr>              <dbl>
## 1 setosa  Sepal.Length         5.1
## 2 setosa  Sepal.Length         4.9
## 3 setosa  Sepal.Length         4.7
## # ... with 597 more rows
```

In this statement, we can see the convenience of dispensing with quotation marks for the new (`part` and `dimension`) and existing (`Species`) column names. Use of bare names as above triggers errors when package code is tested, requiring the use of a less convenient but more consistent and reliable syntax instead. As it is also possible to pass column names as strings but not together with the subtraction operator, equivalent code becomes more verbose but with the intention explicit and easier to grasp.

```
long_iris.tb_1  <-  gather(iris.tb,  key  =  "part",  value  =  "dimension",  setd-
iff(colnames(iris.tb), "Species"))
long_iris.tb_1
## # A tibble: 600 x 3
##    Species part           dimension
##    <fct>   <chr>              <dbl>
## 1 setosa  Sepal.Length         5.1
## 2 setosa  Sepal.Length         4.9
## 3 setosa  Sepal.Length         4.7
## # ... with 597 more rows
```

> ⚠ Altering R's normal interpretation of the name passed as an argument to `key` and `value` prevents these arguments from being recognized as the name of a variable in the calling environment. We need to use a new operator `!!` to restore the normal R behavior.
>
> ```
> part <- "not part"
> long_iris.tb_2 <- gather(iris.tb, key = !!part, value = dimension, -Species)
> long_iris.tb_2
> ## # A tibble: 600 x 3
> ##    Species `not part`   dimension
> ##    <fct>   <chr>            <dbl>
> ## 1 setosa  Sepal.Length       5.1
> ## 2 setosa  Sepal.Length       4.9
> ## 3 setosa  Sepal.Length       4.7
> ## # ... with 597 more rows
> ```
>
> This syntax has been recently subject to debate and led to John Mount developing package 'seplyr' which provides wrappers on functions and methods from 'dplyr' that respect standard evaluation (SE). At the time of writing, 'seplyr' can be considered as experimental.

> To better understand why I added –Species as an argument, edit the code by removing it, and execute the statement to see how the returned tibble is different.

For the reverse operation, converting from long form to wide form, we use spread().

```
spread(long_iris.tb, key = c(!!part, Species), value = dimension) # does not work!!
```

> Starting from version 1.0.0 of 'tidyr', gather() and spread() are deprecated and replaced by pivot_longer() and pivot_wider(). These new functions use a different syntax but are not yet fully stable.

6.7 Data manipulation with 'dplyr'

> The first advantage a user of the 'dplyr' functions and methods sees is the completeness of the set of operations supported and the symmetry and consistency among the different functions. A second advantage is that almost all the functions are defined not only for objects of class tibble, but also for objects of class data.table (packages 'dtplyr') and for SQL databases ('dbplyr'), with consistent syntax (see also section 8.14 on page 325). A further variant exists in package 'seplyr', supporting a different syntax stemming from the use of "standard evaluation" (SE) instead of non-standard evaluation (NSE). A downside of 'dplyr' and much of the 'tidyverse' is that the syntax is not yet fully stable. Additionally, some function and method names either override those in base R or clash with names used in other packages. R itself is extremely stable and expected to remain forward and backward compatible for a long time. For code intended to remain in use for years, the fewer packages it depends on, the less maintenance it will need. When using the 'tidyverse' we need to be prepared to revise our own dependent code after any major revision to the 'tidyverse' packages we may use.

> A new package, 'poorman', implements many of the same words and grammar as 'dplyr' using pure R in the implementation instead of compiled C++

and C code. This light-weight approach could be useful when dealing with relatively small data sets or when the use of R's data frames instead of tibbles is preferred.

6.7.1 Row-wise manipulations

Assuming that the data is stored in long form, row-wise operations are operations combining values from the same observation event—i.e., calculations within a single row of a data frame or tibble. Using functions `mutate()` and `transmute()` we can obtain derived quantities by combining different variables, or variables and constants, or applying a mathematical transformation. We add new variables (columns) retaining existing ones using `mutate()` or we assemble a new tibble containing only the columns we explicitly specify using `transmute()`.

Different from usual R syntax, with `tibble()`, `mutate()` and `transmute()` we can use values passed as arguments, in the statements computing the values passed as later arguments. In many cases, this allows more concise and easier to understand code.

```
tibble(a = 1:5, b = 2 * a)
## # A tibble: 5 x 2
##        a     b
##    <int> <dbl>
## 1     1     2
## 2     2     4
## 3     3     6
## # ... with 2 more rows
```

Continuing with the example from the previous section, we most likely would like to split the values in variable `part` into `plant_part` and `part_dim`. We use `mutate()` from 'dplyr' and `str_extract()` from 'stringr'. We use regular expressions as arguments passed to `pattern`. We do not show it here, but `mutate()` can be used with variables of any `mode`, and calculations can involve values from several columns. It is even possible to operate on values applying a lag or, in other words, using rows displaced relative to the current one.

```
long_iris.tb %.>%
  mutate(.,
         plant_part = str_extract(part, "^[:alpha:]*"),
         part_dim = str_extract(part, "[:alpha:]*$")) -> long_iris.tb
long_iris.tb
## # A tibble: 600 x 5
##    Species part            dimension plant_part part_dim
##    <fct>   <chr>               <dbl> <chr>      <chr>
## 1 setosa  Sepal.Length          5.1 Sepal      Length
## 2 setosa  Sepal.Length          4.9 Sepal      Length
## 3 setosa  Sepal.Length          4.7 Sepal      Length
## # ... with 597 more rows
```

In the next few chunks, we print the returned values rather than saving them in variables. In normal use, one would combine these functions into a pipe using operator %.>% (see section 6.5 on page 187).

Function arrange() is used for sorting the rows—makes sorting a data frame or tibble simpler than by using sort() and order(). Here we sort the tibble long_iris.tb based on the values in three of its columns.

```
arrange(long_iris.tb, Species, plant_part, part_dim)
## # A tibble: 600 x 5
##    Species part          dimension plant_part part_dim
##    <fct>   <chr>             <dbl> <chr>      <chr>
## 1 setosa  Petal.Length        1.4 Petal      Length
## 2 setosa  Petal.Length        1.4 Petal      Length
## 3 setosa  Petal.Length        1.3 Petal      Length
## # ... with 597 more rows
```

Function filter() can be used to extract a subset of rows—similar to subset() but with a syntax consistent with that of other functions in the 'tidyverse'. In this case, 300 out of the original 600 rows are retained.

```
filter(long_iris.tb, plant_part == "Petal")
## # A tibble: 300 x 5
##    Species part          dimension plant_part part_dim
##    <fct>   <chr>             <dbl> <chr>      <chr>
## 1 setosa  Petal.Length        1.4 Petal      Length
## 2 setosa  Petal.Length        1.4 Petal      Length
## 3 setosa  Petal.Length        1.3 Petal      Length
## # ... with 297 more rows
```

Function slice() can be used to extract a subset of rows based on their positions—an operation that in base R would use positional (numeric) indexes with the [,] operator: long_iris.tb[1:5,].

```
slice(long_iris.tb, 1:5)
## # A tibble: 5 x 5
##    Species part          dimension plant_part part_dim
##    <fct>   <chr>             <dbl> <chr>      <chr>
## 1 setosa  Sepal.Length        5.1 Sepal      Length
## 2 setosa  Sepal.Length        4.9 Sepal      Length
## 3 setosa  Sepal.Length        4.7 Sepal      Length
## # ... with 2 more rows
```

Function select() can be used to extract a subset of columns-this would be done with positional (numeric) indexes with [,] in base R, passing them to the second argument as numeric indexes or column names in a vector. Negative indexes in base R can only be numeric, while select() accepts bare column names prepended with a minus for exclusion.

```
select(long_iris.tb, -part)
## # A tibble: 600 x 4
##    Species dimension plant_part part_dim
##    <fct>       <dbl> <chr>      <chr>
## 1 setosa        5.1 Sepal      Length
## 2 setosa        4.9 Sepal      Length
## 3 setosa        4.7 Sepal      Length
## # ... with 597 more rows
```

In addition, `select()` as other functions in 'dplyr' accept "selectors" returned by functions `starts_with()`, `ends_with()`, `contains()`, and `matches()` to extract or retain columns. For this example we use the "wide"-shaped `iris.tb` instead of `long_iris.tb`.

```
select(iris.tb, -starts_with("Sepal"))
## # A tibble: 150 x 3
##    Petal.Length Petal.Width Species
##           <dbl>       <dbl> <fct>
## 1           1.4         0.2 setosa
## 2           1.4         0.2 setosa
## 3           1.3         0.2 setosa
## # ... with 147 more rows
```

```
select(iris.tb, Species, matches("pal"))
## # A tibble: 150 x 3
##    Species Sepal.Length Sepal.Width
##    <fct>          <dbl>       <dbl>
## 1 setosa           5.1         3.5
## 2 setosa           4.9         3
## 3 setosa           4.7         3.2
## # ... with 147 more rows
```

Function `rename()` can be used to rename columns, whereas base R requires the use of both `names()` and `names<-()` and *ad hoc* code to match new and old names. As shown below, the syntax for each column name to be changed is `<new name> = <old name>`. The two names can be given either as bare names as below or as character strings.

```
rename(long_iris.tb, dim = dimension)
## # A tibble: 600 x 5
##    Species part                 dim plant_part part_dim
##    <fct>   <chr>              <dbl> <chr>      <chr>
## 1 setosa  Sepal.Length         5.1 Sepal      Length
## 2 setosa  Sepal.Length         4.9 Sepal      Length
## 3 setosa  Sepal.Length         4.7 Sepal      Length
## # ... with 597 more rows
```

6.7.2 Group-wise manipulations

Another important operation is to summarize quantities by groups of rows. Contrary to base R, the grammar of data manipulation, splits this operation in two: the setting of the grouping, and the calculation of summaries. This simplifies the code, making it more easily understandable when using pipes compared to the approach of base R `aggregate()`, and it also makes it easier to summarize several columns in a single operation.

⚠ It is important to be aware that grouping is persistent, and may also affect other operations on the same data frame or tibble if it is saved or piped and reused. Grouping is invisible to users except for its side effects and because of this can lead to erroneous and surprising results from calculations. Do not save grouped tibbles or data frames, and always make sure that inputs and outputs, at the head and tail of a pipe, are not grouped, by using `ungroup()` when needed.

The first step is to use `group_by()` to "tag" a tibble with the grouping. We create a *tibble* and then convert it into a *grouped tibble*. Once we have a grouped tibble, function `summarise()` will recognize the grouping and use it when the summary values are calculated.

```
tibble(numbers = 1:9, letters = rep(letters[1:3], 3)) %.>%
  group_by(., letters) %.>%
  summarise(.,
            mean_numbers = mean(numbers),
            median_numbers = median(numbers),
            n = n())
## # A tibble: 3 x 4
##    letters mean_numbers median_numbers     n
## *  <chr>          <dbl>          <int> <int>
## 1 a                  4              4     3
## 2 b                  5              5     3
## 3 c                  6              6     3
```

⚠ How is grouping implemented for data frames and tibbles? In our case as our tibble belongs to class `tibble_df`, grouping adds `grouped_df` as the most derived class. It also adds several attributes with the grouping information in a format suitable for fast selection of group members. To demonstrate this, we need to make an exception to our recommendation above and save a grouped tibble to a variable.

```
my.tb <- tibble(numbers = 1:9, letters = rep(letters[1:3], 3))
is.grouped_df(my.tb)
## [1] FALSE

class(my.tb)
## [1] "tbl_df"     "tbl"          "data.frame"

names(attributes(my.tb))
## [1] "names"      "row.names" "class"


my_gr.tb <- group_by(.data = my.tb, letters)
is.grouped_df(my_gr.tb)
## [1] TRUE

class(my_gr.tb)
## [1] "grouped_df" "tbl_df"       "tbl"          "data.frame"
```

```
names(attributes(my_gr.tb))
## [1] "names"     "row.names" "groups"     "class"

setdiff(attributes(my_gr.tb), attributes(my.tb))
## [[1]]
## # A tibble: 3 x 2
##    letters         .rows
## * <chr>      <list<int>>
## 1 a                  [3]
## 2 b                  [3]
## 3 c                  [3]
##
## [[2]]
## [1] "grouped_df" "tbl_df"     "tbl"          "data.frame"


my_ugr.tb <- ungroup(my_gr.tb)
class(my_ugr.tb)
## [1] "tbl_df"     "tbl"          "data.frame"

names(attributes(my_ugr.tb))
## [1] "names"     "row.names" "class"


all(my.tb == my_gr.tb)
## [1] TRUE

all(my.tb == my_ugr.tb)
## [1] TRUE

identical(my.tb, my_gr.tb)
## [1] FALSE

identical(my.tb, my_ugr.tb)
## [1] TRUE
```

The tests above show that members are in all cases the same as operator == tests for equality at each position in the tibble but not the attributes, while attributes, including `class` differ between normal tibbles and grouped ones and so they are not *identical* objects.

If we replace `tibble` by `data.frame` in the first statement, and rerun the chunk, the result of the last statement in the chunk is FALSE instead of TRUE. At the time of writing starting with a `data.frame` object, applying grouping with `group_by()` followed by ungrouping with `ungroup()` has the side effect of converting the data frame into a tibble. This is something to be very much aware of, as there are differences in how the extraction operator [,] behaves in the two cases. The safe way to write code making use of functions from 'dplyr' and 'tidyr' is to always use tibbles instead of data frames.

6.7.3 Joins

Joins allow us to combine two data sources which share some variables. Variables in common are used to match the corresponding rows before "joining" variables (i.e., columns) from both sources together. There are several *join* functions in 'dplyr'. They differ mainly in how they handle rows that do not have a match between data sources.

We create here some artificial data to demonstrate the use of these functions. We will create two small tibbles, with one column in common and one mismatched row in each.

```
first.tb <- tibble(idx = c(1:4, 5), values1 = "a")
second.tb <- tibble(idx = c(1:4, 6), values2 = "b")
```

Below we apply the functions exported by 'dplyr': full_join(), left_join(), right_join() and inner_join(). These functions always retain all columns, and in case of multiple matches, keep a row for each matching combination of rows. We repeat each example with the arguments passed to x and y swapped to more clearly show their different behavior.

A full join retains all unmatched rows filling missing values with NA. By default the match is done on columns with the same name in x and y, but this can be changed by passing an argument to parameter by. Using by one can base the match on columns that have different names in x and y, or prevent matching of columns with the same name in x and y (example at end of the section).

```
full_join(x = first.tb, y = second.tb)

## Joining, by = "idx"
## # A tibble: 6 x 3
##     idx values1 values2
## * <dbl> <chr>   <chr>
## 1     1 a       b
## 2     2 a       b
## 3     3 a       b
## 4     4 a       b
## 5     5 a       <NA>
## 6     6 <NA>    b
```

```
full_join(x = second.tb, y = first.tb)

## Joining, by = "idx"
## # A tibble: 6 x 3
##     idx values2 values1
## * <dbl> <chr>   <chr>
## 1     1 b       a
## 2     2 b       a
## 3     3 b       a
## 4     4 b       a
## 5     6 b       <NA>
## 6     5 <NA>    a
```

Left and right joins retain rows not matched from only one of the two data sources, x and y, respectively.

```
left_join(x = first.tb, y = second.tb)

## Joining, by = "idx"
## # A tibble: 5 x 3
##     idx values1 values2
## * <dbl> <chr>   <chr>
## 1     1 a       b
## 2     2 a       b
## 3     3 a       b
## 4     4 a       b
## 5     5 a       <NA>


left_join(x = second.tb, y = first.tb)

## Joining, by = "idx"
## # A tibble: 5 x 3
##     idx values2 values1
## * <dbl> <chr>   <chr>
## 1     1 b       a
## 2     2 b       a
## 3     3 b       a
## 4     4 b       a
## 5     6 b       <NA>


right_join(x = first.tb, y = second.tb)

## Joining, by = "idx"
## # A tibble: 5 x 3
##     idx values1 values2
## * <dbl> <chr>   <chr>
## 1     1 a       b
## 2     2 a       b
## 3     3 a       b
## 4     4 a       b
## 5     6 <NA>    b


right_join(x = second.tb, y = first.tb)

## Joining, by = "idx"
## # A tibble: 5 x 3
##     idx values2 values1
## * <dbl> <chr>   <chr>
## 1     1 b       a
## 2     2 b       a
## 3     3 b       a
## 4     4 b       a
## 5     5 <NA>    a
```

An inner join discards all rows in x that do not have a matching row in y and *vice versa*.

```
inner_join(x = first.tb, y = second.tb)

## Joining, by = "idx"
```

```
## # A tibble: 4 x 3
##     idx values1 values2
## * <dbl> <chr>   <chr>
## 1     1 a       b
## 2     2 a       b
## 3     3 a       b
## 4     4 a       b
```

```
inner_join(x = second.tb, y = first.tb)
```

```
## Joining, by = "idx"
## # A tibble: 4 x 3
##     idx values2 values1
## * <dbl> <chr>   <chr>
## 1     1 b       a
## 2     2 b       a
## 3     3 b       a
## 4     4 b       a
```

Next we apply the *filtering join* functions exported by 'dplyr': `semi_join()` and `anti_join()`. These functions only return a tibble that always contains only the columns from x, but retains rows based on their match to rows in y.

A semi join retains rows from x that have a match in y.

```
semi_join(x = first.tb, y = second.tb)
```

```
## Joining, by = "idx"
## # A tibble: 4 x 2
##     idx values1
##   <dbl> <chr>
## 1     1 a
## 2     2 a
## 3     3 a
## 4     4 a
```

```
semi_join(x = second.tb, y = first.tb)
```

```
## Joining, by = "idx"
## # A tibble: 4 x 2
##     idx values2
##   <dbl> <chr>
## 1     1 b
## 2     2 b
## 3     3 b
## 4     4 b
```

A anti-join retains rows from x that do not have a match in y.

```
anti_join(x = first.tb, y = second.tb)
```

```
## Joining, by = "idx"
## # A tibble: 1 x 2
##     idx values1
##   <dbl> <chr>
## 1     5 a
```

```
anti_join(x = second.tb, y = first.tb)

## Joining, by = "idx"
## # A tibble: 1 x 2
##     idx values2
##   <dbl> <chr>
## 1     6 b
```

We here rename column idx in first.tb to demonstrate the use of by to specify which columns should be searched for matches.

```
first2.tb <- rename(first.tb, idx2 = idx)
full_join(x = first2.tb, y = second.tb, by = c("idx2" = "idx"))
## # A tibble: 6 x 3
##    idx2 values1 values2
## * <dbl> <chr>   <chr>
## 1     1 a       b
## 2     2 a       b
## 3     3 a       b
## 4     4 a       b
## 5     5 a       <NA>
## 6     6 <NA>    b
```

6.8   Further reading

An in-depth discussion of the 'tidyverse' is outside the scope of this book. Several books describe in detail the use of these packages. As several of them are under active development, recent editions of books such as *R for Data Science* (Wickham and Grolemund 2017) are the most useful.

7

Grammar of graphics

The commonality between science and art is in trying to see profoundly—to develop strategies of seeing and showing.

Edward Tufte's answer to Charlotte Thralls
An Interview with Edward R. Tufte, 2004

7.1 Aims of this chapter

Three main data plotting systems are available to R users: base R, package 'lattice' (Sarkar 2008) and package 'ggplot2' (Wickham and Sievert 2016), the last one being the most recent and currently most popular system available in R for plotting data. Even two different sets of graphics primitives (i.e., those used to produce the simplest graphical elements such as lines and symbols) are available in R, those in base R and a newer one in the 'grid' package (Murrell 2011).

In this chapter you will learn the concepts of the grammar of graphics, on which package 'ggplot2' is based. You will also learn how to build several types of data plots with package 'ggplot2'. As a consequence of the popularity and flexibility of 'ggplot2', many contributed packages extending its functionality have been developed and deposited in public repositories. However, I will focus mainly on package 'ggplot2' only briefly describing a few of these extensions.

7.2 Packages used in this chapter

If the packages used in this chapter are not yet installed in your computer, you can install them as shown below, as long as package 'learnrbook' is already installed.

```
install.packages(learnrbook::pkgs_ch_ggplot)
```

To run the examples included in this chapter, you need first to load some packages from the library (see section 5.2 on page 163 for details on the use of packages).

```
library(learnrbook)
library(wrapr)
library(scales)
library(ggplot2)
library(ggrepel)
library(gginnards)
library(ggpmisc)
library(ggbeeswarm)
library(ggforce)
library(tikzDevice)
library(lubridate)
library(tidyverse)
library(patchwork)
```

7.3 Introduction to the grammar of graphics

What separates 'ggplot2' from base R and trellis/lattice plotting functions is the use of a grammar of graphics (the reason behind 'gg' in the name of package 'ggplot2'). What is meant by grammar in this case is that plots are assembled piece by piece using different "nouns" and "verbs" (Cleveland 1985). Instead of using a single function with many arguments, plots are assembled by combining different elements with operators + and %+%. Furthermore, the construction is mostly semantics-based and to a large extent, how plots look when printed, displayed, or exported to a bitmap or vector-graphics file is controlled by themes.

We can think of plotting as representing the observations or data in a graphical language. We use the properties of graphical objects to represent different aspects of our data. An observation can consist of multiple recorded values. Say an observation of air temperature may be defined by a position in 3-dimensional space and a point in time, in addition to the temperature itself. An observation for the size and shape of a plant can consist of height, stem diameter, number of leaves, size of individual leaves, length of roots, fresh mass, dry mass, etc. If we are interested in the relationship between height and stem diameter, we may want to use cartesian coordinates, *mapping* stem diameter to the x dimension of the plot and the height to the y dimension. The observations could be represented on the plot by points and/or joined by lines.

The grammar of graphics allows us to design plots by combining various elements in ways that are nearly orthogonal. In other words, the majority of the possible combinations of "words" yield valid plots as long as we assemble them respecting the rules of the grammar. This flexibility makes 'ggplot2' extremely powerful as we can build plots and even types of plots which were not even considered while designing the 'ggplot2' package.

When a plot is built, the whole plot and its components are created as R objects that can be saved in the workspace or written to a file as objects. The graphical representation is generated when the object is printed, explicitly or automatically. The same ggplot object can be rendered into different bitmap and vector graphic formats for display or printing.

Even if we do not explicitly add them all, default elements may be used. The production of a rendered graphic with package 'ggplot2' can be represented as a flow of information: data → scale → statistic → aesthetic → geometry → coordinate → ggplot → theme → rendered graphic.

7.3.1 Data

The data to be plotted must be available as a data.frame or tibble, with data stored so that each row represents a single observation event, and the columns are different values observed in that single event. In other words, in long form (so-called "tidy data") as described in chapter 6. The variables to be plotted can be numeric, factor, character, and time or date stored as POSIXct.

7.3.2 Mapping

When we design a plot, we need to map data variables to aesthetics (or graphic properties). Most plots will have an x dimension, which is considered an *aesthetic*, and a variable containing numbers mapped to it. The position on a 2D plot of, say, a point, will be determined by x and y aesthetics, while in a 3D plot, three aesthetics need to be mapped x, y and z. Many aesthetics are not related to coordinates, they are properties, like color, size, shape, line type, or even rotation angle, which add an additional dimension on which to represent the values of variables and/or constants.

7.3.3 Geometries

Geometries are "words" that describe the graphics representation of the data: for example, geom_point(), plots a point or symbol for each observation, while geom_line(), draws line segments between observations. Some geometries rely by default on statistics, but most "geoms" default to the identity statistics. Each time a *geometry* is used to add a graphical representation of data to a plot, we say that a new *layer* has been added. The name *layer* reflects the fact that each new layer added is plotted on top of the layers already present in the plot, or rather when a plot is printed the layers will be generated in the order they were added to the ggplot object. For example, one layer in a plot can display the observations, another layer a regression line fitted to them, and a third one may contain annotations such an equation or a text label.

7.3.4 Statistics

Statistics are "words" that represent calculation of summaries or some other operation on the values from the data. When *statistics* are used for a computation, the returned value is passed directly to a *geometry*, and consequently adding an *statistics* also adds a layer to the plot. For example, stat_smooth() fits a smoother, and stat_summary() applies a summary function. Statistics are applied automatically by group when data have been grouped by mapping additional aesthetics such as color to a factor.

7.3.5   Scales

Scales give the "translation" or mapping between data values and the aesthetic values to be actually plotted. Mapping a variable to the "color" aesthetic (also recognized when spelled as "colour") only tells that different values stored in the mapped variable will be represented by different colors. A scale, such as `scale_color_continuous()`, will determine which color in the plot corresponds to which value in the variable. Scales can also define transformations on the data, which are used when mapping data values to aesthetic values. All continuous scales support transformations—e.g., in the case of x and y aesthetics, positions on the plotting region or viewport will be affected by the transformation, while the original values will be used for tick labels along the axes. Scales are used for all aesthetics, including continuous variables, such as numbers, and categorical ones such as factors. The grammar of graphics allows only one scale per *aesthetic* and plot. This restriction is imposed by design to avoid ambiguity (e.g., it ensures that the red color will have the same "meaning" in all plot layers where the `color` *aesthetic* is mapped to data). Scales have limits with observations falling outside these limits being ignored (replaced by NA) rather than passed to statistics or geometries—it is easy to unintentionally drop observations when setting scale limits manually as warning messages report that NA values have been omitted.

7.3.6   Coordinate systems

The most frequently used coordinate system when plotting data, the cartesian system, is the default for most *geometries*. In the cartesian system, x and y are represented as distances on two orthogonal (at 90°) axes. Additional coordinate systems are available in 'ggplot2' and through extensions. For example, in the polar system of coordinates, the x values are mapped to angles around a central point and y values to the radius. Another example is the ternary system of coordinates, an extension of the grammar implemented in package 'ggtern', that allows the construction of ternary plots. Setting limits to a coordinate system changes the region of the plotting space visible in the plot, but does not discard observations. In other words, when using *statistics*, observations located outside the coordinate limits, i.e., not visible in the rendered plot, will still be included in computations.

7.3.7   Themes

How the plots look when displayed or printed can be altered by means of themes. A plot can be saved without adding a theme and then printed or displayed using different themes. Also, individual theme elements can be changed, and whole new themes defined. This adds a lot of flexibility and helps in the separation of the data representation aspects from those related to the graphical design.

7.3.8 Plot construction

We have described above the components of the grammar of graphics: aesthetics (`aes`), for example color, geometric elements `geom_`... such as lines and points, statistics `stat_`..., scales `scale_`..., coordinate systems `coord_`..., and themes `theme_`.... In this section we will see how plots are assembled from these elements.

As the workings and use of the grammar are easier to show by example than to explain with words, will show how to build plots of increasing complexity. All elements of a plot have defaults, although in some cases these defaults result in empty plots. Defaults make it possible to create a plot very succinctly. We use function `ggplot()` to create the skeleton for a plot, which can be enhanced, but also printed as is.

```
ggplot()
```

The plot above is of little use without any data, so we next pass a data frame object, in this case `mtcars`—`mtcars` is a data set included in R; to learn more about this data set, type `help("mtcars")` at the R command prompt.

```
ggplot(data = mtcars)
```

Once the data are available, we need to *map* the quantities in the data onto graphical features in the plot, or *aesthetics*. When plotting in two dimensions, we need to map variables in the data to at least the x and y aesthetics. This mapping can be seen in the chunk below by its effect on the plotting area ranges that now match the ranges of the mapped variables, extended by a small margin. The axis labels also reflect the names of the mapped variables, however, there is no graphical element yet displayed for the individual observations.

```
ggplot(data = mtcars,
       aes(x = disp, y = mpg))
```

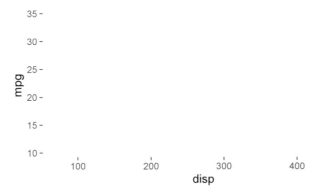

To make observations visible in the plot we need to add a suitable *geometry* or geom to the plot. Here we display the observations as points using geom_point()

```
ggplot(data = mtcars,
       aes(x = disp, y = mpg)) +
  geom_point()
```

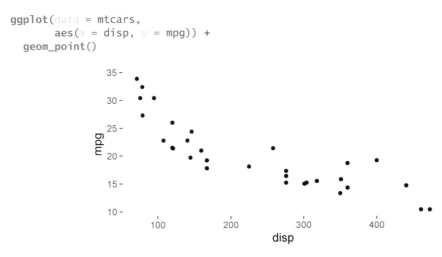

⚠ In the examples above, the plots were printed automatically, which is the default at the R console. However, as with other R objects, ggplots can be assigned to a variable,

```
p <- ggplot(data = mtcars,
            aes(x = disp, y = mpg)) +
       geom_point()
```

and printed at a later time.

```
print(p)
```

📊💻 Above we have seen how to build a plot, layer by layer, using the grammar of graphics. We have also seen how to save a ggplot. We can peep into the innards of this object using summary().

```
summary(p)
```

We can view the structure of the ggplot object with `str()`.

Package 'gginnards' provides methods `str()`, `num_layers()`, `top_layer()` and `mapped_vars()`. Use these methods to explore ggplot objects with different numbers of layers or mappings. You will see that the plot elements that were added to the plot are stored as members of a list with nested lists forming a tree-like structure.

Although *aesthetics* can be mapped to variables in the data, they can also be set to constant values, but only within layers, not as whole-plot defaults.

```
ggplot(data = mtcars,
       aes(x = disp, y = mpg)) +
  geom_point(color = "red", shape = "square")
```

While a geometry directly constructs a graphical representation of the observations in the data, a *statistics* or `stat` "sits" in-between the data and a `geom`, applying some computation, usually but not always, to produce a statistical summary of the data. Here we add a fitted line using `stat_smooth()` with its output added to the plot using `geom_line()` passed by name with `"line"` as an argument to `stat_smooth`. We fit a linear regression, using `lm()` as the method.

```
ggplot(data = mtcars,
       aes(x = disp, y = mpg)) +
  geom_point() +
  stat_smooth(geom = "line", method = "lm", formula = y ~ x)
```

We haven't yet added some of the elements of the grammar described above: *scales, coordinates* and *themes*. The plots were rendered anyway because these elements have defaults which are used when we do not set them explicitly. We next will see examples in which they are explicitly set. We start with a scale using a logarithmic transformation. This works like plotting by hand using graph paper with rulings spaced according to a logarithmic scale. Tick marks continue to be expressed in the original units, but statistics are applied to the transformed data. In other words, a transformed scale affects the values before they are passed to *statistics*, and the linear regression will be fitted to log10() transformed y values and the original x values.

```
ggplot(data = mtcars,
       aes(x = disp, y = mpg)) +
  geom_point() +
  stat_smooth(geom = "line", method = "lm", formula = y ~ x) +
  scale_y_log10()
```

The range limits of a scale can be set manually, instead of automatically by default. These limits create a virtual *window into the data*: observations outside the scale limits remain hidden and are not mapped to aesthetics—i.e., these observations are not included in the graphical representation or used in calculations. Crucially, when using *statistics* the computations are only applied to observations that fall within the limits of all scales in use. These limits *indirectly* affect the plotting area when the plotting area is automatically set based on the range of the (within limits) data—even the mapping to values of a different aesthetics may change when a subset of the data are selected by manually setting the limits of a scale.

In contrast to *scale limits, coordinates* function as a *zoomed view* into the plotting area, and do not affect which observations are visible to *statistics*. The coordinate system, as expected, is also determined by this grammar element—here we use cartesian coordinates which are the default, but we manually set y limits.

```
ggplot(data = mtcars,
       aes(x = disp, y = mpg)) +
  geom_point() +
  stat_smooth(geom = "line", method = "lm", formula = y ~ x) +
  coord_cartesian(ylim = c(15, 25))
```

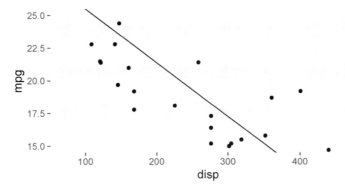

The next example uses a coordinate system transformation. When the transformation is applied to the coordinate system, it affects only the plotting—it sits between the geom and the rendering of the plot. The transformation is applied to the values returned by any *statistics*. The straight line fitted is plotted on the transformed coordinates as a curve, because the model was fitted to the untransformed data and this fitted model is automatically used to obtain the predicted values, which are then plotted after the transformation is applied to them. We have here described only cartesian coordinate systems while other coordinate systems are described in sections 7.4.6 and 7.9 on pages 228 and 272, respectively.

```
ggplot(data = mtcars,
       aes(x = disp, y = mpg)) +
  geom_point() +
  stat_smooth(geom = "line", method = "lm", formula = y ~ x) +
  coord_trans(y = "log10")
```

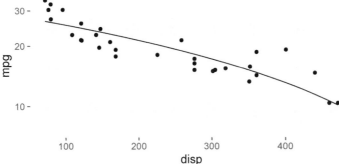

Themes affect the rendering of plots at the time of printing—they can be thought of as style sheets defining the graphic design. A complete theme can override the default gray theme. The plot is the same, the observations are represented in the same way, the limits of the axes are the same and all text is the same. On the other, hand how these elements are rendered by different themes can be drastically different.

```
ggplot(data = mtcars,
       aes(x = disp, y = mpg)) +
  geom_point() +
  theme_classic()
```

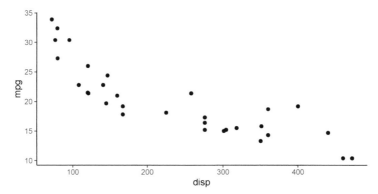

We can also override the base font size and font family. This affects the size of all text elements, as their size is defined relative to the base size. Here we add the same theme as used in the previous example, but with a different base point size for text.

```
ggplot(data = mtcars,
       aes(x = disp, y = mpg)) +
  geom_point() +
  theme_classic(base_size = 20, base_family = "serif")
```

The details of how to set axis labels, tick positions and tick labels will be discussed in depth in section 7.7. Meanwhile, we will use function `labs()` which is a convenience function allowing us to easily set the title and subtitle of a plot and to replace the default `name` of scales used for axis labels—by default `name` is set to the name of the mapped variable. When setting the `name` of scales with `labs()`, we use as parameter names the names of aesthetics and pass as an argument a character string, or an R expression. Here we use `x` and `y`, the names of the two *aesthetics* to which we have mapped two variables in `data`, `disp` and `mpg`.

```
ggplot(data = mtcars,
       aes(x = disp, y = mpg)) +
  geom_point() +
  labs(x = "Engine displacement (cubic inches)",
       y = "Fuel use efficiency\n(miles per gallon)",
       title = "Motor Trend Car Road Tests",
       subtitle = "Source: 1974 Motor Trend US magazine")
```

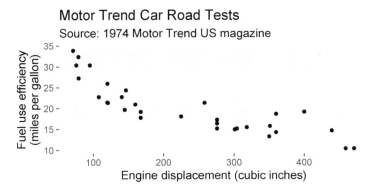

As elsewhere in R, when a value is expected, either a value stored in a variable or statement returning a suitable value can be passed as an argument to be mapped to an *aesthetic.* In other words, the values to be plotted do not need to be stored as variables (or columns) in the data frame passed as an argument to parameter `data`, they can also be computed from these variables. Here we plot miles-per-gallon, `mpg` on the engine displacement per cylinder by dividing `disp` by `cyl` within the call to `aes()`.

```
ggplot(data = mtcars, aes(x = disp / cyl, y = mpg)) +
  geom_point()
```

Each of the elements of the grammar exemplified above has several different members, and many of the individual *geometries* and *statistics* accept arguments that can be used to modify their behavior. There are also more *aesthetics* than those shown above. Multiple data objects as well as multiple mappings can coexist within a single ggplot object. Packages and user code can define new *geometries*, *statistics*, *coordinates* and even implement new *aesthetics.* Individual elements in a theme can also be modified and new complete themes created, re-used and shared. We will describe in the remaining sections of this chapter how to use the grammar of graphics to construct other types of graphical presentations including more complex plots than those in the examples above.

7.3.9   Plots as R objects

We can manipulate ggplot objects and their components in the same way as other R objects. We can operate on them using the operators and methods defined for the "gg" class they belong to. We start by saving a ggplot into a variable.

```
p <- ggplot(data = mtcars,
       aes(x = disp, y = mpg)) +
   geom_point()
```

> ⚠ The separation of plot construction and rendering is possible, because "gg" objects are self-contained. Most importantly, a copy of the data object passed as argument is saved within the plot object. In the example above, p by itself could be saved to a file on disk and loaded into a clean R session, even on another computer, and rendered as long as package 'ggplot2' and its dependencies are available. Another consequence of a copy of the data being stored in the plot object, is that editing the data used to create a "gg" object after its creation does *not* affect rendered plots unless we recreate the "gg" object.
>
> With str() we can explore the structure of any R object, including those of class "gg". We use max.level = 1 to reduce the length of output, but to see deeper into the nested list you can increase the value passed as an argument to max.level or simply accept its default.
>
> ```
> str(p, max.level = 1)
> ```

When we used in the previous section operator + to assemble the plots, we were operating on "anonymous" R objects. In the same way, we can operate on saved or "named" objects.

```
p +
   stat_smooth(geom = "line", method = "lm", formula = y ~ x)
```

> 📊 Reproduce the examples in the previous section, using p defined above as a basis instead of building each plot from scratch.

> ℹ️ In the examples above we have been adding elements one by one, using the + operator. It is also possible to add multiple components in a single operation using a list. This is useful, when we want to save sets of components in a variable so as to reuse them in multiple plots. This saves typing, ensures consistency and can make alterations to a set of similar plots much easier.
>
> ```
> my.layers <- list(
>   stat_smooth(geom = "line", method = "lm", formula = y ~ x),
>   scale_x_log10())
> ```
>
> ```
> p + my.layers
> ```
>

7.3.10 Data and mappings

In the case of simple plots, based on data contained in a single data frame, the usual style is to code a plot as described above, passing an argument, mtcars in these examples, to the data parameter of ggplot(). Data passed in this way becomes the default for all layers in the plot. The same applies to the argument passed to mapping.

```
ggplot(data = mtcars,
       mapping = aes(x = disp, y = mpg)) +
  geom_point()
```

However, the grammar of graphics contemplates the possibility of data and mappings restricted to individual layers. In this case, those passed as arguments to ggplot(), if present, are overridden by arguments passed to individual layers, making it possible to code the same plot as follows.

```
ggplot() +
  geom_point(data = mtcars,
             mapping = aes(x = disp, y = mpg))
```

The default mapping can also be added directly with the + operator, instead of being passed as an argument to `ggplot()`.

```
ggplot(data = mtcars) +
  aes(x = disp, y = mpg) +
  geom_point()
```

It is even possible to have a default mapping for the whole plot, but no default data.

```
ggplot() +
  aes(x = disp, y = mpg) +
  geom_point(data = mtcars)
```

In these examples, the plot remains unchanged, but this flexibility in the grammar allows, in plots containing multiple layers, for each layer to use different data or a different mapping.

📟 The argument passed to parameter `data` of a layer function, can be a function instead of a data frame, if the plot contains default data. In this case, the function is applied to the default data and must return a data frame containing data to be used in the layer.

```
ggplot(data = mtcars,
       mapping = aes(x = disp, y = mpg)) +
  geom_point(size = 4) +
  geom_point(data = function(x){subset(x, cyl == 4)}, color = "yellow",
             size = 1.5)
```

The plot default data can also be operated upon using the 'magritrr' pipe operator, but not the dot-pipe operator from 'wrapr' (see section 6.5 on page 187).

```
ggplot(data = mtcars,
       mapping = aes(x = disp, y = mpg)) +
  geom_point(size = 4) +
  geom_point(data = . %.>% subset(x = ., cyl == 4), color = "yellow",
             size = 1.5)
```

7.4 Geometries

Different geometries support different *aesthetics*. While `geom_point()` supports shape, and `geom_line()` supports linetype, both support x, y, color and size. In this section we will describe the different geometries available in package 'ggplot2' and some examples from packages that extend 'ggplot2'. The graphic output from most code examples will not be shown, with the expectation that readers will run them to see the plots.

Mainly for historical reasons, *geometries* accept a *statistic* as an argument, in the same way as *statistics* accept a *geometry* as an argument. In this section we will only describe *geometries* which have as a default *statistic* stat_identity which passes values directly as mapped. The *geometries* that have other *statistics* as default are described in section 7.5.2 together with the corresponding *statistics*.

7.4.1 Point

As shown earlier in this chapter, geom_point(), can be used to add a layer with observations represented by "points" or symbols. Variable cyl describes the numbers of cylinders in the engines of the cars. It is a numeric variable, and when mapped to color, a continuous color scale is used to represent this variable.

The first examples build scatter plots, because numeric variables are mapped to both x and y. Some scales, like those for color, exist in two "flavors," one suitable for numeric variables (continuous) and another for factors (discrete).

```
ggplot(data = mtcars,
       aes(x = disp, y = mpg, color = cyl)) +
  geom_point()
```

If we convert cyl into a factor, a discrete color scale is used instead of a continuous one.

```
ggplot(data = mtcars,
       aes(x = disp, y = mpg, color = factor(cyl))) +
  geom_point()
```

If we convert cyl into an ordered factor, a different discrete color scale is used by default.

```
ggplot(data = mtcars,
       aes(x = disp, y = mpg, color = ordered(cyl))) +
  geom_point()
```

> 📚 Try a different mapping: `disp` → `color`, `cyl` → `x`. Continue by using `help(mtcars)` and/or `names(mtcars)` to see what variables are available, and then try the combinations that trigger your curiosity—i.e., explore the data.

The mapping between data values and aesthetic values is controlled by scales. Different color scales, and even palettes within a given scale, provide different mappings between data values and rendered colours.

```
ggplot(data = mtcars,
       aes(x = disp, y = mpg, color = factor(cyl))) +
  geom_point() +
  scale_color_brewer(type = "qual", palette = 2)
```

The data, aesthetics mappings, and geometries are the same as in earlier code; to alter how the plot looks, we have changed only the scale and palette used for the color aesthetic. Conceptually it is still exactly the same plot we created earlier, except for the colours used. This is a very important point to understand, because it allows us to separate two different concerns: the semantic structure and the graphic design.

> 📚 Try the different palettes available through the brewer scale. You can play directly with the palettes using function `brewer_pal()` from package 'scales' together with `show_col()`.
>
> ```
> show_col(brewer_pal()(3))
> show_col(brewer_pal(type = "qual", palette = 2, direction = 1)(3))
> ```
>
> Once you have found a suitable palette for these data, redo the plot above with the chosen palette.

When not relying on colors, the most common way of distinguishing groups of observations in scatter plots is to use the `shape` of the points as an *aesthetic*. We need to change a single "word" in the code statement to achieve this different mapping.

```
ggplot(data = mtcars, aes(x = disp, y = mpg, shape = factor(cyl))) +
  geom_point()
```

We can use `scale_shape_manual` to choose each shape to be used. We set three "open" shapes that we will see later are very useful as they obey both `color` and `fill` *aesthetics*.

```
ggplot(data = mtcars, aes(x = disp, y = mpg, shape = factor(cyl))) +
  geom_point() +
  scale_shape_manual(values = c(21, 22, 23))
```

It is also possible to use characters as shapes. The character is centered on the position of the observation. As the numbers used as symbols are self-explanatory, we suppress the default guide or key.

```
ggplot(data = mtcars, aes(x = disp, y = mpg, shape = factor(cyl))) +
  geom_point(size = 2.5) +
  scale_shape_manual(values = c("4", "6", "8"), guide = FALSE)
```

> ⓘ One variable in the data can be mapped to more than one aesthetic, allowing redundant aesthetics. This may seem wasteful, but it is extremely useful as it allows one to produce figures that, even when produced in color, can still be read if reproduced as black-and-white images.
>
> ```
> ggplot(data = mtcars, aes(x = disp, y = mpg,
>                           shape = factor(cyl),
>                           color = factor(cyl))) +
>   geom_point()
> ```

Dot plots are similar to scatter plots but a factor is mapped to either the x or y *aesthetic*. Dot plots are prone to have overlapping observations, and one way of making these points visible is to make them partly transparent by setting a constant value smaller than one for the alpha *aesthetic*.

```
ggplot(data = mtcars, aes(x = factor(cyl), y = mpg)) +
  geom_point(alpha = 1/3)
```

Instead of making the points semitransparent, we can randomly displace them to avoid overlaps. This is called *jitter*, and can be added using `position_jitter()` and the amount of jitter set with `width` as a fraction of the distance between adjacent factor levels in the plot.

```
ggplot(data = mtcars, aes(x = factor(cyl), y = mpg)) +
  geom_point(position = position_jitter(width = 0.05))
```

We can create a "bubble" plot by mapping the `size` *aesthetic* to a continuous variable. In this case, one has to think what is visually more meaningful. Although the radius of the shape is frequently mapped, due to how human perception works, mapping a variable to the area of the shape is more useful by being perceptually closer to a linear mapping. For this example we add a new variable to the plot. The weight of the car in tons and map it to the area of the points.

```
ggplot(data = mtcars, aes(x = disp, y = mpg,
                          color = factor(cyl),
                          size = wt)) +
  scale_size_area() +
  geom_point()
```

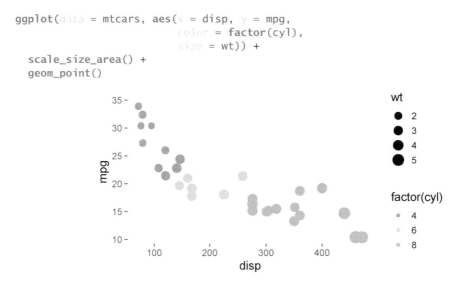

If we use a radius-based scale the "impression" is different.

```
ggplot(data = mtcars, aes(x = disp, y = mpg,
                          color = factor(cyl),
                          size = wt)) +
  scale_size() +
  geom_point()
```

Make the plot, look at it carefully. Check the numerical values of some of the weights, and assess if your perception of the plot matches the numbers behind it.

As a final example summarizing the use of `geom_point()`, we combine different *aesthetics* and *scales* in the same scatter plot.

```
ggplot(data = mtcars, aes(x = disp, y = mpg,
                          shape = factor(cyl),
                          fill = factor(cyl),
                          size = wt)) +
  geom_point(alpha = 0.33, color = "black") +
  scale_size_area() +
  scale_shape_manual(values = c(21, 22, 23))
```

> Play with the code in the chunk above. Remove or change each of the mappings and the scale, display the new plot, and compare it to the one above. Continue playing with the code until you are sure you understand what graphical element in the plot is added or modified by each individual argument or "word" in the code statement.

It is common to draw error bars together with points representing means or medians of observations and `geom_pointrange()` achieves this task based on the values mapped to the `x`, `y`, `ymin` and `ymax`, using `y` for the position of the point and `ymin` and `ymax` for the positions of the ends of the line segment representing a range. Two other *geometries*, `geom_range()` and `geom_errorbar` draw only a segment or a segment with capped ends. They are frequently used together with *statistics* when summaries are calculated on the fly, but can also be used directly when the data summaries are stored in a data frame passed as an argument to `data`.

7.4.2   Rug

Rarely, rug plots are used by themselves. Instead they are usually an addition to scatter plots. An example of the use of `geom_rug()` follows. They make it easier to see the distribution of observations along the *x*- and *y*-axes.

```
ggplot(data = mtcars,
       aes(x = disp, y = mpg, color = factor(cyl))) +
  geom_point() +
  geom_rug()
```

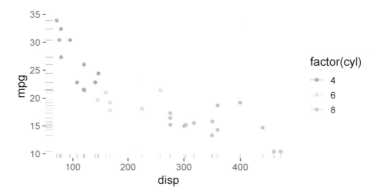

⚠ Rug plots are most useful when the local density of observations is not too high, otherwise rugs become too cluttered and the "rug threads" may overlap. When overlap is moderate, making the segments semitransparent by setting the `alpha` aesthetic to a constant value smaller than one, can make the variation in density easier to appreciate. When the number of observations is large, marginal density plots should be preferred.

7.4.3   Line and area

For line plots we use `geom_line()`. The `size` of a line is its thickness, and as we had `shape` for points, we have `linetype` for lines. In a line plot, observations in successive rows of the data frame, or the subset corresponding to a group, are joined by straight lines. We use a different data set included in R, `Orange`, with data on the growth of five orange trees. See the help page for `Orange` for details.

```
ggplot(data = Orange,
       aes(x = age, y = circumference, linetype = Tree)) +
  geom_line()
```

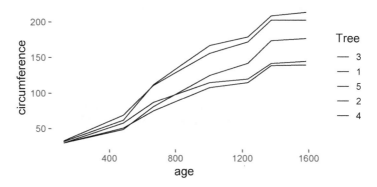

Instead of drawing a line joining the successive observations, we may want to draw a disconnected straight-line segment for each observation or row in the data. In this case, we use `geom_segment()` which accepts x, xend, y and yend as mapped aesthetics. `geom_curve()` draws curved lines, and the curvature, control points, and angles can be controlled through additional *aesthetics*. These two *geometries* support arrow heads at their ends. Other *geometries* useful for drawing lines or segments are geom_path(), which is similar to geom_line(), but instead of joining observations according to the values mapped to x, it joins them according to their row-order in data, and geom_spoke(), which is similar to geom_segment() but using a polar parametrization, based on x, y for origin, and angle and radius for the segment. Finally, geom_step() plots only vertical and horizontal lines to join the observations, creating a stepped line.

```
ggplot(data = Orange,
       aes(x = age, y = circumference, linetype = Tree)) +
  geom_step()
```

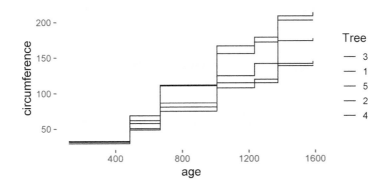

Using the following toy data, make three plots using geom_line(), geom_path(), and geom_step to add a layer.

```
toy.df <- data.frame(x = c(1,3,2,4), y = c(0,1,0,1))
```

While geom_line() draws a line joining observations, geom_area() supports filling the area below the line according to the fill *aesthetic*. In contrast geom_ribbon draws two lines based on the x, ymin and ymax *aesthetics*, with the space between the lines filled according to the fill *aesthetic*. Finally, geom_polygom is similar to geom_path() but connects the extreme observations forming a closed polygon that supports fill.

Much of what was described above for geom_point can be adapted to geom_line, geom_ribbon, geom_area and other *geometries* described in this section. In some cases, it is useful to stack the areas—e.g., when the values represent parts of a bigger whole. In the next, contrived, example, we stack the growth of the different trees by using position = "stack" instead of the default position = "identity". (Compare the y axis of the figure below to that drawn using geom_line on page 222.)

```
ggplot(data = Orange,
       aes(x = age, y = circumference, fill = Tree)) +
  geom_area(position = "stack")
```

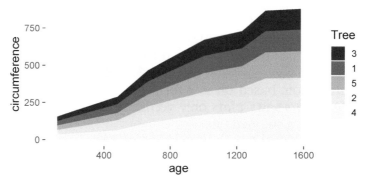

Finally, three *geometries* for drawing lines across the whole plotting area: `geom_hline`, `geom_vline` and `geom_abline`. The first two draw horizontal and vertical lines, respectively, while the third one draws straight lines according to the *aesthetics* `slope` and `intercept` determining the position. The lines drawn with these three geoms extend to the edge of the plotting area.

`geom_hline` and `geom_vline` require a single aesthetic, `yintercept` and `xintercept`, respectively. Different from other geoms, the data for these aesthetics can also be passed as constant numeric vectors. The reason for this is that these geoms are most frequently used to annotate plots rather than plotting observations. Let's assume that we want to highlight an event at the age of 1000 days.

```
ggplot(data = Orange,
       aes(x = age, y = circumference, fill = Tree)) +
  geom_area(position = "stack") +
  geom_vline(xintercept = 1000, color = "gray75") +
  geom_vline(xintercept = 1000, linetype = "dotted")
```

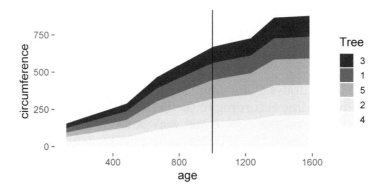

📊 Change the order of the three layers in the example above. How did the figure change? What order is best? Would the same order be the best for a scatter plot? And would it be necessary to add two `geom_vline()` layers?

7.4.4 Column

The *geometry* `geom_col()` can be used to create *column plots* where each bar represents an observation or case in the data.

> ⚠️ R users not familiar yet with 'ggplot2' are frequently surprised by the default behavior of `geom_bar()` as it uses `stat_count()` to produce a histogram, rather than plotting values as is (see section 7.5.4 on page 245). `geom_col()` is identical to `geom_bar()` but with `"identity"` as the default statistic.

We create artificial data that we will reuse in multiple variations of the next figure.

```
set.seed(654321)
my.col.data <- data.frame(treatment = factor(rep(c("A", "B", "C"), 2)),
                          group = factor(rep(c("male", "female"), c(3, 3))),
                          measurement = rnorm(6) + c(5.5, 5, 7))
```

First we plot data for females only, using defaults for all *aesthetics* except *x* and *y* which we explicitly map to variables.

```
ggplot(subset(my.col.data, group == "female"),
       aes(x = treatment, y = measurement)) +
  geom_col()
```

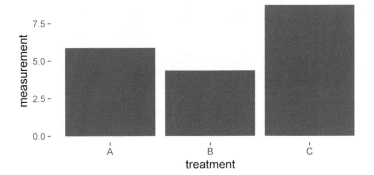

We play with *aesthetics* to produce a plot with a semi-formal style—e.g., suitable for a science popularization article or book. See section 7.7 and section 7.10 for information on scales and themes, respectively. We set `width = 0.5` to make the bars narrower. Setting `color = "white"` overrides the default color of the lines bordering the bars.

```
ggplot(my.col.data, aes(x = treatment, y = measurement, fill = group)) +
  geom_col(color = "white", width = 0.5) +
  scale_fill_grey() + theme_dark()
```

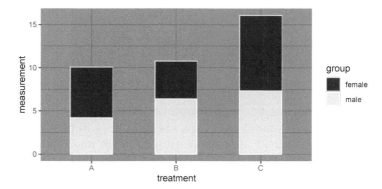

We next use a formal style, and in addition, put the bars side by side by setting `position = "dodge"` to override the default `position = "stack"`. Setting `color = NA` removes the lines bordering the bars.

```
ggplot(my.col.data, aes(x = treatment, y = measurement, fill = group)) +
    geom_col(color = NA, position = "dodge") +
    scale_fill_grey() + theme_classic()
```

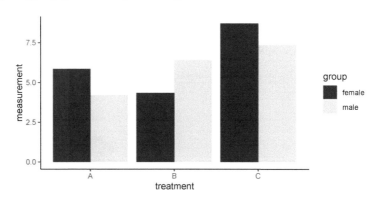

Change the argument to `position`, or let the default be active, until you understand its effect on the figure. What is the difference between *positions* `"identity"`, `"dodge"` and `"stack"`?

Use constants as arguments for *aesthetics* or map variable `treatment` to one or more of the *aesthetics* used by `geom_col()`, such as `color`, `fill`, `linetype`, `size`, `alpha` and `width`.

7.4.5 Tiles

We can draw square or rectangular tiles with `geom_tile()` producing tile plots or simple heat maps.

We here generate 100 random draws from the F distribution with degrees of freedom $v_1 = 5, v_2 = 20$.

```
set.seed(1234)
randomf.df <- data.frame(F.value = rf(100, df1 = 5, df2 = 20),
                         x = rep(letters[1:10], 10),
                         y = LETTERS[rep(1:10, rep(10, 10))])
```

geom_tile() requires aesthetics x and y, with no defaults, and width and height with defaults that make all tiles of equal size filling the plotting area.

```
ggplot(randomf.df, aes(x, y, fill = F.value)) +
  geom_tile()
```

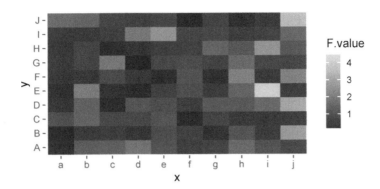

We can set color = "gray75" and size = 1 to make the tile borders more visible as in the example below, or use a contrasting color, to better delineate the borders of the tiles. What to use will depend on whether the individual tiles add meaningful information. In cases like when rows of tiles correspond to individual genes and columns to discrete treatments, the use of contrasting tile borders is preferable. In contrast, in the case when the tiles are an approximation to a continuous surface such as measurements on a regular spatial grid, it is best to suppress the tile borders.

```
ggplot(randomf.df, aes(x, y, fill = F.value)) +
  geom_tile(color = "gray75", size = 1.33)
```

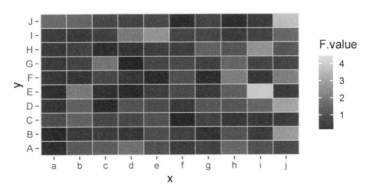

📊 Play with the arguments passed to parameters color and size in the example above, considering what features of the data are most clearly perceived in each of the plots you create.

Any continuous fill scale can be used to control the appearance. Here we show a tile plot using a gray gradient, with missing values in red.

```
ggplot(randomf.df, aes(x, y, fill = F.value) +
  geom_tile(color = "white") +
  scale_fill_gradient(low = "gray15", high = "gray85", na.value = "red")
```

In contrast to `geom_tile()`, `geom_rect()` draws rectangular tiles based on the position of the corners, mapped to aesthetics `xmin`, `xmax`, `ymin` and `ymax`.

7.4.6 Simple features (sf)

'ggplot2' version 3.0.0 or later supports the plotting of shape data similar to the plotting in geographic information systems (GIS) through `geom_sf()` and its companions, `geom_sf_text()`, `geom_sf_label()`, and `stat_sf()`. This makes it possible to display data on maps, for example, using different fill values for different regions. Special *coordinate* `coord_sf()` can be used to select different projections for maps. The *aesthetic* used is called `geometry` and contrary to all the other aesthetics we have seen until now, the values to be mapped are of class `sfc` containing *simple features* data with multiple components. Manipulation of simple features data is supported by package 'sf'. This subject exceeds the scope of this book, so a single and very simple example follows.

```
nc <- sf::st_read(system.file("shape/nc.shp", package = "sf"), quiet = TRUE)
ggplot(nc) +
  geom_sf(aes(fill = AREA), color = "gray90")
```

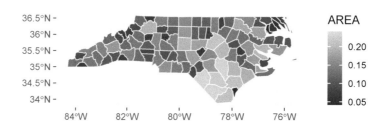

7.4.7 Text

We can use `geom_text()` or `geom_label()` to add text labels to observations. For `geom_text()` and `geom_label()`, the aesthetic `label` provides the text to be plotted and the usual aesthetics `x` and `y`, the location of the labels. As one would expect, the `color` and `size` aesthetics can also be used for the text.

```
ggplot(data = mtcars, aes(x = disp, y = mpg,
                    color = factor(cyl),
                    size = wt,
                    label = cyl)) +
  scale_size() +
```

```
geom_point() +
geom_text(color = "darkblue", size = 3)
```

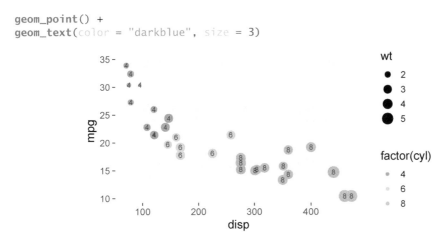

In addition, `angle` and `vjust` and `hjust` can be used to rotate the text and adjust its position. The default value of 0.5 for both `hjust` and `vjust` sets the center of the text at the supplied x and y coordinates. "Vertical" and "horizontal" for justification refer to the text, not the plot. This is important when `angle` is different from zero. Values larger than 0.5 shift the label left or down, and values smaller than 0.5, right or up with respect to its x and y coordinates. A value of 1 or 0 sets the text so that its edge is at the supplied coordinate. Values outside the range 0 … 1 shift the text even farther away, however, still using units based on the length or height of the text label. Recent versions of 'ggplot2' make possible justification using character constants for alignment: `"left"`, `"middle"`, `"right"`, `"bottom"`, `"center"` and `"top"`, and two special alignments, `"inward"` and `"outward"`, that automatically vary based on the position in the plotting area.

In the case of `geom_label()` the text is enclosed in a box, which obeys the `fill` *aesthetic* and takes additional parameters (described starting at page 231) allowing control of the shape and size of the box. However, `geom_label()` does not support rotation with the `angle` aesthetic.

⚠ You should be aware that R and 'ggplot2' support the use of UNICODE, such as UTF8 character encodings in strings. If your editor or IDE supports their use, then you can type Greek letters and simple maths symbols directly, and they *may* show correctly in labels if a suitable font is loaded and an extended encoding like UTF8 is in use by the operating system. Even if UTF8 is in use, text is not fully portable unless the same font is available, as even if the character positions are standardized for many languages, most UNICODE fonts support at most a small number of languages. In principle one can use this mechanism to have labels both using other alphabets and languages like Chinese with their numerous symbols mixed in the same figure. Furthermore, the support for fonts and consequently character sets in R is output-device dependent. The font encoding used by R by default depends on the default locale settings of the operating system, which can also lead to garbage printed to the console or wrong characters being plotted running the same code on a different computer from the one where a script was created. Not all is lost, though, as

> R can be coerced to use system fonts and Google fonts with functions provided by packages 'showtext' and 'extrafont'. Encoding-related problems, especially in MS-Windows, are common.

In the remaining examples, with output not shown, we use `geom_text` or `geom_label` together with `geom_point` as this is how they may be used to label observations.

```
my.data <-
  data.frame(x = 1:5,
             y = rep(2, 5),
             label = c("a", "b", "c", "d", "e"))

ggplot(my.data, aes(x, y, label = label)) +
  geom_text(angle = 45, hjust = 1.5, size = 8) +
  geom_point()
```

> ☷ Modify the example above to use `geom_label()` instead of `geom_text()` using, in addition, the `fill` aesthetic.

In the next example we select a different font family, using the same characters in the Roman alphabet. The names `"sans"` (the default), `"serif"` and `"mono"` are recognized by all graphics devices on all operating systems. Additional fonts are available for specific graphic devices, such as the 35 "PDF" fonts by the `pdf()` device. In this case, their names can be queried with `names(pdfFonts())`.

```
ggplot(my.data, aes(x, y, label = label)) +
  geom_text(angle = 45, hjust = 1.5, size = 8, family = "serif") +
  geom_point()
```

> ☷ In the examples above the character strings were all of the same length, containing a single character. Redo the plots above with longer character strings of various lengths mapped to the `label` *aesthetic*. Do also play with justification of these labels.

Plotting (mathematical) expressions involves mapping to the `label` aesthetic character strings that can be parsed as expressions, and setting `parse = TRUE` (see section 7.12 on page 282). Here, we build the character strings using `paste()` but, of course, they could also have been entered one by one. This use of `paste()` provides an example of recycling of shorter vectors (see section 2.10 on page 45).

```
my.data <-
  data.frame(x = 1:5, y = rep(2, 5), label = paste("alpha[", 1:5, "]", sep = ""))
my.data$label
## [1] "alpha[1]" "alpha[2]" "alpha[3]" "alpha[4]" "alpha[5]"
```

Text and labels do not automatically expand the plotting area past their anchoring coordinates. In the example above, we need to use expand_limits() to ensure that the text is not clipped at the edge of the plotting area.

```
ggplot(my.data, aes(x, y, label = label)) +
  geom_text(hjust = -0.2, parse = TRUE, size = 6) +
  geom_point() +
  expand_limits(x = 5.2)
```

In the example above, we mapped to label the text to be parsed. It is also possible, and usually preferable, to build suitable labels on the fly within aes() when setting the mapping for label. Here we use geom_text() with strings to be parsed into expressions created on the fly within the call to aes(). The same approach can be used for regular character strings not requiring parsing.

```
ggplot(my.data, aes(x, y, label = paste("alpha[", x, "]", sep = ""))) +
  geom_text(hjust = -0.2, parse = TRUE, size = 6) +
  geom_point()
```

As geom_label() obeys the same parameters as geom_text() except for angle, we briefly describe below only the additional parameters compared to geom_text(). We may want to alter the default width of the border line or the color used to fill the rectangle, or to change the "roundness" of the corners. To suppress the border line, use label.size = 0. Corner roundness is controlled by parameter label.r and the size of the margin around the text by label.padding.

```
my.data <-
  data.frame(x = 1:5, y = rep(2, 5),
             label = c("one", "two", "three", "four", "five"))

ggplot(my.data, aes(x, y, label = label)) +
  geom_label(hjust = -0.2, size = 6,
             label.size = 0L,
             label.r = unit(0, "lines"),
             label.padding = unit(0.15, "lines"),
             fill = "yellow", alpha = 0.5) +
  geom_point() +
  expand_limits(x = 5.6)
```

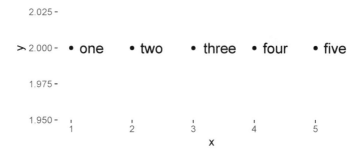

> 🔖 Play with the arguments to the different parameters and with the *aesthetics* to get an idea of what can be done with them. For example, use thicker border lines and increase the padding so that a visually well-balanced margin is retained. You may also try mapping the `fill` and `color` *aesthetics* to factors in the data.

If the parameter `check_overlap` of `geom_text()` is set to TRUE, text overlap will be avoided by suppressing the text that would otherwise overlap other text. *Repulsive* versions of `geom_text()` and `geom_label()`, `geom_text_repel()` and `geom_label_repel()`, are available in package 'ggrepel'. These *geometries* avoid overlaps by automatically repositioning the text or labels. Please read the package documentation for details of how to control the repulsion strength and direction, and the properties of the segments linking the labels to the position of their data coordinates. Nearly all aesthetics supported by `geom_text()` and `geom_label()` are supported by the repulsive versions. However, given that a segment connects the label or text to its anchor point, several properties of these segments can also be controlled with aesthetics or arguments.

```
ggplot(data = mtcars,
       aes(x = disp, y = mpg, color = factor(cyl), size = wt, label = cyl)) +
  scale_size() +
  geom_point(alpha = 1/3) +
  geom_text_repel(color = "black", size = 3,
                  min.segment.length = 0.2, point.padding = 0.1)
```

7.4.8 Plot insets

The support for insets in 'ggplot2' is confined to annotation_custom() which was designed to be used for static annotations expected to be the same in each panel of a plot (the use of annotations is described in section 7.8). Package 'ggpmisc' provides geoms that mimic geom_text() in relation to the *aesthetics* used, but that similarly to geom_sf(), expect that the column in data mapped to the label aesthetics are lists of objects containing multiple pieces of information, rather than atomic vectors. Similar to geom_sf() these geoms do not inherit the plot's default mappings to aesthetics. Three geometries are currently available: geom_table(), geom_plot() and geom_grob().

⚠ Given that geom_table(), geom_plot() and geom_grob() will rarely use a mapping inherited from the whole plot, by default they do not inherit it. Either the mapping should be supplied as an argument to these functions or their parameter inherit.aes explicitly set to TRUE.

The plotting of tables by mapping a list of data frames to the label *aesthetic* is done with geom_table. Positioning, justification, and angle work as for geom_text and are applied to the whole table. Only tibble objects (see documentation of package 'tibble') can contain, as variables, lists of data frames, so this *geometry* requires the use of tibble objects to store the data. The table(s) are created as 'grid' grob objects, collected in a tree and added to the ggplot object as a new layer.

We first generate a tibble containing summaries from the data, formatted as character strings, wrap this tibble in a list, and store this list as a column in another tibble. To accomplish this, we use functions from the 'tidyverse' described in chapter 6.

```
mtcars %.>%
  group_by(., cyl) %.>%
  summarize(.,
            "mean wt" = format(mean(wt), digits = 2),
            "mean disp" = format(mean(disp), digits = 0),
            "mean mpg" = format(mean(mpg), digits = 0)) -> my.table
table.tb <- tibble(x = 500, y = 35, table.inset = list(my.table))


ggplot(data = mtcars, aes(x = disp, y = mpg,
                          color = factor(cyl),
                          size = wt,
                          label = cyl)) +
  scale_size() +
  geom_point() +
  geom_table(data = table.tb,
             aes(x = x, y = y, label = table.inset),
             color = "black", size = 3)
```

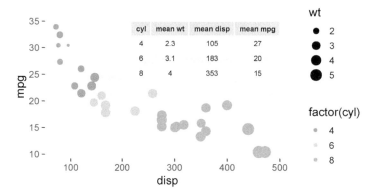

The `color` and `size` aesthetics control the text in the table(s) as a whole. It is also possible to rotate the table(s) using `angle`. As with text labels, justification is interpreted in relation to table-text orientation. We set the `y = 0` in `data.tb` and then use `vjust = 1` to position the top of the table at this coordinate value.

```
ggplot(data = mtcars, aes(x = disp, y = mpg, color = factor(cyl))) +
  geom_point() +
  geom_table(data = table.tb,
             aes(x = x, y = y, label = table.inset),
             color = "blue", size = 3,
             hjust = 1, vjust = 0, angle = 90)
```

Parsed text, using R's *plotmath* syntax is supported in the table, with fallback to plain text in case of parsing errors, on a cell-by-cell basis. We end this section with a simple example, which even if not very useful, demonstrates that `geom_table()` behaves like a "normal" ggplot *geometry* and that a table can be the only layer in a ggplot if desired. The addition of multiple tables with a single call to `geom_table()` by passing a `tibble` with multiple rows as an argument for `data` is also possible.

```
tb.pm <- tibble('x^0' = 1,
                'x^1' = 1:5,
                'x^2' = (1:5)^2,
                'x^3' = (1:5)^3)
data.tb <- tibble(x = 1, y = 1, table.inset = list(tb.pm))
ggplot(data.tb, mapping = aes(x, y, label = table.inset)) +
  geom_table(inherit.aes = TRUE, size = 7, parse = TRUE) +
  theme_void()
```

> ⌨ The *geometry* `geom_table()` uses functions from package 'gridExtra' to build a graphical object for the table. The use of table themes was not yet supported by this geometry at the time of writing.

Geometry `geom_plot()` works much like `geom_table()`, but instead of expecting a list of data frames or tibbles to be mapped to the `label` aesthetics, it expects a list of ggplots (objects of class `gg`). This allows adding as an inset to a ggplot, another ggplot. In the times when plots were hand drafted with India ink on paper, the use of inset plots was more frequent than nowadays. Inset plots can be very useful for zooming-in on parts of a main plot where observations are crowded and for displaying summaries based on the observations shown in the main plot. The inset plots are nested in viewports which control the dimensions of the inset plot, and aesthetics `vp.height` and `vp.width` control their sizes—with defaults of 1/3 of the height and width of the plotting area of the main plot. Themes can be applied separately to the main and inset plots.

In the first example of inset plots, we include one of the summaries shown above as an inset table. We first create a tibble containing the plot to be inset.

```
mtcars %.>%
  group_by(., cyl) %.>%
  summarize(., mean.mpg = mean(mpg)) %.>%
  ggplot(data = .,
         aes(factor(cyl), mean.mpg, fill = factor(cyl))) +
  scale_fill_discrete(guide = FALSE) +
  scale_y_continuous(name = NULL) +
    geom_col() +
    theme_bw(8) -> my.plot
plot.tb <- tibble(x = 500, y = 35, plot.inset = list(my.plot))
```

```
ggplot(data = mtcars, aes(x = disp, y = mpg,
                          color = factor(cyl))) +
  geom_point() +
  geom_plot(data = plot.tb,
            aes(x = x, y = y, label = plot.inset),
            vp.width = 1/2,
            hjust = "inward", vjust = "inward")
```

In the second example we add the zoomed version of the same plot as an inset. 1) Manually set limits to the coordinates to zoom into a region of the main plot, 2) set the *theme* of the inset, 3) remove axis labels as they are the same as in the main plot, 4) and 5) highlight the zoomed-in region in the main plot. This fairly complex example shows how a new extension to 'ggplot2' can integrate well into the grammar of graphics paradigm. In this example, to show an alternative approach, instead of collecting all the data into a data frame, we map constant values directly to the various aesthetics within `annotate()` (see section 7.8 on page 269).

```
p.main <- ggplot(data = mtcars, aes(x = disp, y = mpg, color = factor(cyl))) +
  geom_point()
```

```
p.inset <- p.main +
  coord_cartesian(xlim = c(270, 330), ylim = c(14, 19)) +
  labs(x = NULL, y = NULL) +
  scale_color_discrete(guide = FALSE) +
  theme_bw(8) + theme(aspect.ratio = 1)
p.main +
  geom_plot(x = 480, y = 34, label = list(p.inset), vp.height = 1/2,
            hjust = "inward", vjust = "inward") +
  annotate(geom = "rect", fill = NA, color = "black",
           xmin = 270, xmax = 330, ymin = 14, ymax = 19,
           linetype = "dotted")
```

Geometry `geom_grob()` works much like `geom_table()` and `geom_plot()` but expects a list of 'grid' graphical objects, called `grob` for short. This adds generality at the expense of having to separately create the grobs either using 'grid' or by converting other objects into grobs. This geometry is as flexible as `annotation_custom()` with respect to the grobs, but behaves as a *geometry*. We show an example that adds two bitmaps to the plot. The bitmaps are read from PNG files, converted into grobs, and added to the plot as a new layer. The PNG bitmaps used have a transparent background.

```
file1.name <-
  system.file("extdata", "Isoquercitin.png", package = "ggpmisc", mustwork = TRUE)
Isoquercitin <- magick::image_read(file1.name)
file2.name <-
  system.file("extdata", "Robinin.png", package = "ggpmisc", mustwork = TRUE)
Robinin <- magick::image_read(file2.name)
grob.tb <- tibble(x = c(0, 100), y = c(10, 20), height = 1/3, width = c(1/2),
                  grobs = list(grid::rasterGrob(image = Isoquercitin),
                               grid::rasterGrob(image = Robinin)))

ggplot() +
  geom_grob(data = grob.tb,
            aes(x = x, y = y, label = grobs, vp.height = height, vp.width = width),
            hjust = "inward", vjust = "inward")
```

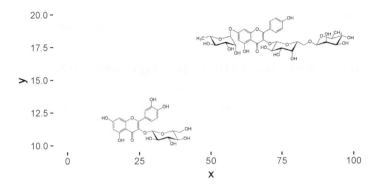

☕ Grid graphics provide the low-level functions that both 'ggplot2' and 'lattice' use under the hood. Grid supports different types of units for expressing the coordinates of positions within the plotting area. All examples outside this text box use "native" data coordinates, however, coordinates can be also given in physical units like "mm". More useful when working with scalable plots is to use "npc" *normalized parent coordinates*, which are expressed as numbers in the range 0 to 1, relative to the dimensions of the sides of the current *viewport*, with origin at the lower left corner.

Package 'ggplot2' interprets x and y coordinates in "native" data coordinates, and trickery seems to be needed to get around this limitation. A rather general solution is provided by package 'ggpmisc' through *aesthetics* npcx and npcy and *geometries* that support them. At the time of writing, geom_text_npc(), geom_label_npc(), geom_table_npc(), geom_plot_npc() and geom_grob_npc(). These *geometries* are useful for annotating plots and adding insets at positions relative to the plotting area that remain always consistent across different plots, or across panels when using facets with free axis limits. Being geometries they provide freedom in the elements added to different panels and their positions.

```
ggplot(data = mtcars, aes(x = disp, y = mpg, color = factor(cyl))) +
  geom_point() +
  geom_label_npc(npcx = 0.5, npcy = 0.9, label = "a label", color = "black")
```

7.5 Statistics

Before learning about 'ggplot2' *statistics*, it is important to have clear how the mapping of factors to *aesthetics* works. When a factor, for example, is mapped to `color`, it creates a new grouping, with the observations matching a given level of the factor, corresponding to a group. Most *statistics* operate on the data for each of these groups separately, returning a summary for each group, for example, the mean of the observations in a group.

7.5.1 Functions

In addition to plotting data from a data frame with variables to map to *x* and *y* aesthetics, it is possible to have only a variable mapped to *x* and use `stat_function()` to compute the values to be mapped to *y* using an R function. This avoids the need to generate data beforehand as even the number of data points to be generated can be set in `geom_function()`. Any R function, user defined or not, can be used as long as it is vectorized, with the length of the returned vector equal to the length of the vector passed as first argument to it. The variable mapped to x determines the range, and the argument to parameter n of `geom_function()` the length of the generated vector that is passed as first argument to fun when it is called to generate the values to be napped to y. These are the *x* and *y* values passed to the *geometry*.

 We start with the Normal distribution function. We rely on the defaults n = 101 and geom = "path".

```
ggplot(data.frame(x = -3:3), aes(x = x)) +
  stat_function(fun = dnorm)
```

Using a list we can even pass by name additional arguments to use when the function is called.

```
ggplot(data.frame(x = -3:3), aes(x = x)) +
  stat_function(fun = dnorm, args = list(mean = 1, sd = .5))
```

> 📘 Edit the code above so as to plot in the same figure three curves, either for three different values for mean or for three different values for sd.

Named user-defined functions (not shown), and anonymous functions (below) can also be used.

```
ggplot(data.frame(x = 0:1), aes(x = x)) +
  stat_function(fun = function(x, a, b){a + b * x^2},
                args = list(a = 1, b = 1.4))
```

> **▲!** Edit the code above to use a different function, such as e^{x+k}, adjusting the argument(s) passed through `args` accordingly. Do this by means of an anonymous function, and by means of an equivalent named function defined by your code.

7.5.2 Summaries

The summaries discussed in this section can be superimposed on raw data plots, or plotted on their own. Beware, that if scale limits are manually set, the summaries will be calculated from the subset of observations within these limits. Scale limits can be altered when explicitly defining a scale or by means of functions `xlim()` and `ylim()`. See section 7.9 on page 272 for an explanation of how coordinate limits can be used to zoom into a plot without excluding of x and y values from the data.

It is possible to summarize data on the fly when plotting. We describe in the same section the calculation of measures of central tendency and of variation, as `stat_summary()` allows them to be calculated simultaneously and added together with a single layer.

For use in the examples, we generate some normally distributed artificial data.

```
fake.data <- data.frame(
  y = c(rnorm(10, mean = 2, sd = 0.5),
        rnorm(10, mean = 4, sd = 0.7)),
  group = factor(c(rep("A", 10), rep("B", 10)))
)
```

We will reuse a "base" scatter plot in a series of examples, so that the differences are easier to appreciate. We first add just the mean. In this case, we need to pass as an argument to `stat_summary()`, the `geom` to use, as the default one, `geom_pointrange()`, expects data for plotting error bars in addition to the mean. This example uses a hyphen character as the constant value of `shape` (see the example for `geom_point()` on page 219 on the use of digits as `shape`). Instead of passing `"mean"` as an argument to parameter `fun` (earlier called `fun.y`), we can pass, if desired, other summary functions like `"median"`. In the case of these functions that return a single computed value, we pass them, or character strings with their names, as an argument to parameter `fun`.

```
ggplot(data = fake.data, aes(y = y, x = group)) +
  geom_point(shape = 21) +
  stat_summary(fun = "mean", geom = "point",
               color = "red", shape = "-", size = 10)
```

To pass as an argument a function that returns a central value like the mean plus confidence or other limits, we use parameter `fun.data` instead of `fun`. In the next example we add means and confidence intervals for $p = 0.95$ (the default) assuming normality.

```
stat_summary(fun.data = "mean_cl_normal", color = "red", size = 1, alpha = 0.7)
```

We can override the default of $p = 0.95$ for confidence intervals by setting, for example, `conf.int = 0.90` in the list of arguments passed to the function. The intervals can also be computed without assuming normality, using the empirical distribution estimated from the data by bootstrap. To achieve this we pass to `fun.data` the argument `"mean_cl_boot"` instead of `"mean_cl_normal"`.

```
stat_summary(fun.data = "mean_cl_boot",
             fun.args = list(conf.int = 0.90),
             color = "red", size = 1, alpha = 0.7)
```

For $\bar{x} \pm$ s.e. we should pass `"mean_se"` and for $\bar{x} \pm$ s.d. `"mean_sdl"`.

```
stat_summary(fun.data = "mean_se",
             color = "red", size = 1, alpha = 0.7)
```

We do not give an example here, but it is possible to use user-defined functions instead of the functions exported by package 'ggplot2' (based on those in package 'Hmisc'). Because arguments to the function used, except for the first one containing the variable in `data` mapped to the y aesthetic, are supplied as a named list through parameter `fun.args`, the names used for parameters in the function definition need only match the names in this list.

Finally, we plot the means in a scatter plot, with the observations superimposed on the error bars as a result of the order in which the layers are added to the plot. In this case, we set `fill`, `color` and `alpha` (transparency) to constants, but in more complex data sets, mapping them to factors in `data` can be used for grouping of observations. Here, adding two plot layers with `stat_summary()` allows us to plot the mean and the error bars using different colors.

```
ggplot(data = fake.data, aes(y = y, x = group)) +
  stat_summary(fun = "mean", geom = "point",
               fill = "white", color = "black") +
  stat_summary(fun.data = "mean_cl_boot",
```

```
              geom = "errorbar",
              width = 0.1, size = 1, color = "red") +
 geom_point(size = 3, alpha = 0.3)
```

We can plot means, or other summaries, by group mapped to x (class in this example) as columns by passing "col" as an argument to geom. In this way we avoid the need to compute the summaries in advance.

```
ggplot(mpg, aes(class, hwy)) +
  stat_summary(geom = "col", fun = mean)
```

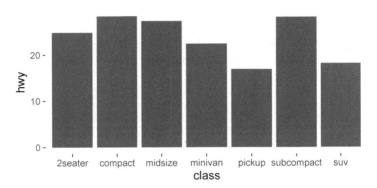

We can easily add error bars to the column plot. We use size to make the lines of the error bars thicker. The default *geometry* in stat_summary() is geom_pointrange(), so we can pass "linerange" as an argument for geom to eliminate the point.

```
 stat_summary(geom = "linerange", fun.data = "mean_se",
              size = 1, color = "red")
```

Passing "errorbar" instead of "linerange" to geom results in traditional "capped" error bars. However, this type of error bar has been criticized as adding unnecessary clutter to plots (Tufte 1983). We can use width to reduce the width of the caps at the ends of the error bars.

If we have already calculated values for the summaries, we can still obtain the same plots by mapping variables to the *aesthetics* required by geom_errorbar() and geom_linerange(): x, y, ymax and ymin.

☕ The "reverse" syntax is also valid, as we can add the *geometry* to the plot object and pass the *statistics* as an argument to it. In general in this book we avoid this alternative syntax for the sake of consistency.

```
ggplot(mpg, aes(class, hwy)) +
  geom_col(stat = "summary", fun = mean)
```

7.5.3   Smoothers and models

The *statistic* stat_smooth() fits a smooth curve to observations in the case when the scales for *x* and *y* are continuous—the corresponding *geometry* geom_smooth() uses this *statistic*, and differs only in how arguments are passed to formal parameters. For the first example, we use stat_smooth() with the default smoother, a spline. The type of spline is automatically chosen based on the number of observations and informed by a message. The formula must be stated using the names of the *x* and *y* aesthetics, rather the names of the mapped variables in mtcars.

```
ggplot(data = mtcars, aes(x = disp, y = mpg)) +
      stat_smooth(formula = y ~ x)
```

In most cases we will want to plot the observations as points together with the smoother. We can plot the observation on top of the smoother, as done here, or the smoother on top of the observations.

```
ggplot(data = mtcars, aes(x = disp, y = mpg)) +
  stat_smooth(formula = y ~ x) +
  geom_point()
```

```
## `geom_smooth()` using method = 'loess'
```

Instead of using the default spline, we can fit a different model. In this example we use a linear model as smoother, fitted by lm().

```
    stat_smooth(method = "lm", formula = y ~ x) +
```

These data are really grouped, so we map variable cyl to the color *aesthetic.* Now we get three groups of points with different colours but also three separate smooth lines.

```
ggplot(data = mtcars, aes(x = disp, y = mpg, color = factor(cyl))) +
  stat_smooth(method = "lm", formula = y ~ x) +
  geom_point()
```

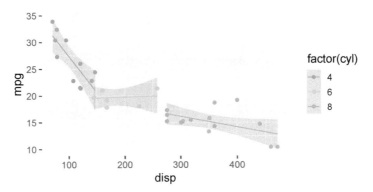

To obtain a single smoother for the three groups, we need to set the mapping of the color *aesthetic* to a constant within `stat_smooth`. This local value overrides the default `color` mapping set in `ggplot()` just for this plot layer. We use `"black"` but this could be replaced by any other color definition known to R.

```
ggplot(data = mtcars, aes(x = disp, y = mpg, color = factor(cyl))) +
  stat_smooth(method = "lm", formula = y ~ x, color = "black") +
  geom_point()
```

Instead of using the `formula` for a linear regression as smoother, we pass a different `formula` as an argument. In this example we use a polynomial of order 2.

```
ggplot(data = mtcars, aes(x = disp, y = mpg, color = factor(cyl))) +
  stat_smooth(method = "lm", formula = y ~ poly(x, 2), color = "black") +
  geom_point()
```

It is possible to use other types of models, including GAM and GLM, as smoothers, but we will give only two simple examples of the use of `nls()` to fit a model non-linear in its parameters (see section 4.8 on page 140 for details about fitting this same model with `nls()`). In the first one we fit a Michaelis-Menten equation to reaction rate (`rate`) versus reactant concentration (`conc`). `Puromycin` is a data set included in the R distribution. Function `ssmicmen()` is also from R, and is a *self-starting* implementation of the Michaelis-Menten equation. Thanks to this, even though the fit is done with an iterative algorithm, we do not need to explicitly provide starting values for the parameters to be fitted. We need to set `se = FALSE` because standard errors are not supported by the `predict()` method for `nls` fitted models.

```
ggplot(Puromycin, aes(conc, rate, color = state)) +
  geom_point() +
  geom_smooth(method = "nls",
              formula =  y ~ SSmicmen(x, Vm, K),
              se = FALSE)
```

In the second example we define the same model directly in the model formula, and provide the starting values explicitly. The names used for the parameters to be fitted can be chosen at will, within the restrictions of the R language, but of course the names used in `formula` and `start` must match each other.

```
ggplot(Puromycin, aes(conc, rate, color = state)) +
  geom_point() +
  geom_smooth(method = "nls",
              method.args = list(formula =  y ~ (Vmax * x) / (k + x),
                                 start = list(Vmax = 200, k = 0.05)),
              se = FALSE)
```

In some cases it is desirable to annotate plots with fitted model equations or fitted parameters. One way of achieving this is by fitting the model and then extracting the parameters to manually construct text strings to use for text or label annotations. However, package 'ggpmisc' makes it possible to automate such annotations in many cases.

```
my.formula <- y ~ poly(x, 2)
ggplot(data = mtcars, aes(x = disp, y = mpg, color = factor(cyl))) +
  stat_smooth(method = "lm", formula = my.formula, color = "black") +
  stat_poly_eq(formula = my.formula, aes(label = ..eq.label..),
               color = "black", parse = TRUE, label.x.npc = 0.3) +
  geom_point()
```

This same package makes it possible to annotate plots with summary tables from a model fit.

```
my.formula <- y ~ poly(x, 2)
ggplot(data = mtcars, aes(x = disp, y = mpg, color = factor(cyl))) +
  stat_smooth(method = "lm", formula = my.formula, color = "black") +
  stat_fit_tb(method = "lm",
              method.args = list(formula = my.formula),
              color = "black",
              tb.vars = c(Parameter = "term",
                          Estimate = "estimate",
```

```
                    "s.e." = "std.error",
                    "italic(t)" = "statistic",
                    "italic(P)" = "p.value"),
           label.y.npc = "top", label.x.npc = "right",
           parse = TRUE) +
  geom_point()
```

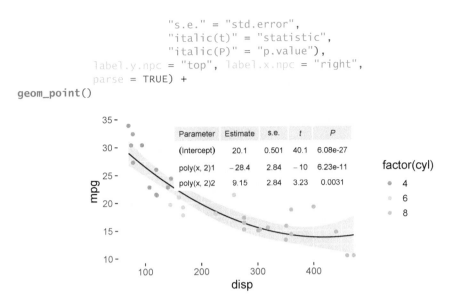

Package 'ggpmisc' provides additional *statistics* for the annotation of plots based on fitted models. Please see the package documentation for details.

7.5.4 Frequencies and counts

When the number of observations is rather small, we can rely on the density of graphical elements to convey the density of the observations. For example, scatter plots using well-chosen values for alpha can give a satisfactory impression of the density. Rug plots, described in section 7.4.2 on page 221, can also satisfactorily convey the density of observations along x and/or y axes. Such approaches do not involve computations, while the *statistics* described in this section do. Frequencies by value-range (or bins) and empirical density functions are summaries especially useful when the number of observations is large. These summaries can be computed in one or more dimensions.

Histograms are defined by how the plotted values are calculated. Although histograms are most frequently plotted as bar plots, many bar or "column" plots are not histograms. Although rarely done in practice, a histogram could be plotted using a different *geometry* using stat_bin(), the *statistic* used by default by geom_histogram(). This *statistic* does binning of observations before computing frequencies, and is suitable for continuous x scales. When a factor is mapped to x, stat_count() should be used, which is the default stat for geom_bar(). These two *geometries* are described in this section about statistics, because they default to using statistics different from stat_identity() and consequently summarize the data.

As before, we generate suitable artificial data.

```
set.seed(12345)
my.data <-
  data.frame(x = rnorm(200),
             y = c(rnorm(100, -1, 1), rnorm(100, 1, 1)),
             group = factor(rep(c("A", "B"), c(100, 100))) )
```

We could have relied on the default number of bins automatically computed by the `stat_bin()` statistic, however, we here set it to 15 with `bins = 15`. It is important to remember that in this case no variable in `data` is mapped onto the `y` *aesthetic*.

```
ggplot(my.data, aes(x)) +
  geom_histogram(bins = 15)
```

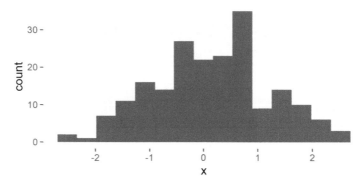

If we create a grouping by mapping a factor to an additional *aesthetic* how the bars created are positioned with respect to each other becomes relevant. We can then plot side by side with `position = "dodge"`, stacked one above the other with `position = "stack"` and overlapping with `position = "identity"` in which case we need to make them semi-transparent with `alpha = 0.5` so that they all remain visible.

```
ggplot(my.data, aes(y, fill = group)) +
  geom_histogram(bins = 15, position = "dodge")
```

The computed values are contained in the `data` that the *geometry* "receives" from the *statistic*. Many statistics compute additional values that are not mapped by default. These can be mapped with `aes()` by enclosing them in a call to `stat()`. From the help page we can learn that in addition to counts in variable `count`, density is returned in variable `density` by this statistic. Consequently, we can create a histogram with the counts per bin expressed as densities whose integral is one (rather than their sum, as the width of the bins is in this case different from one), as follows.

```
ggplot(my.data, aes(y, fill = group)) +
  geom_histogram(mapping = aes(y = stat(density)), bins = 15, position = "dodge")
```

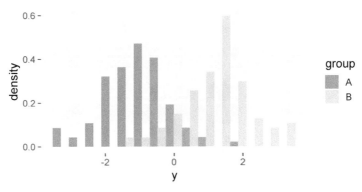

If it were not for the easier to remember name of `geom_histogram()`, adding the layers with `stat_bin()` or `stat_count()` would be preferable as it makes clear that computations on the data are involved.

```
ggplot(my.data, aes(y, fill = group)) +
  stat_bin(bins = 15, position = "dodge")
```

The *statistic* `stat_bin2d`, and its matching *geometry* `geom_bin2d()`, by default compute a frequency histogram in two dimensions, along the x and y *aesthetics*. The frequency for each rectangular tile is mapped onto a `fill` scale. As for `stat_bin()`, density is also computed and available to be mapped as shown above for `geom_histogram`. In this example, to compare dispersion in two dimensions, equal *x* and *y* scales are most suitable, which we achieve by adding `coord_fixed()`, which is a variation of the default `coord_cartesian()` (see section 7.9 on page 272 for details on other systems of coordinates).

```
ggplot(my.data, aes(x, y)) +
  stat_bin2d(bins = 8) +
  coord_fixed(ratio = 1)
```

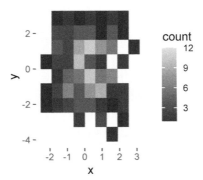

The *statistic* `stat_bin_hex()`, and its matching *geometry* `geom_hex()`, differ from `stat_bin2d()` in their use of hexagonal instead of square tiles. By default the frequency or count for each hexagon is mapped to the `fill` aesthetic, but counts expressed as density are also computed and can be mapped with `aes(fill = stat(density))`.

```
ggplot(my.data, aes(x, y)) +
  stat_bin_hex(bins = 8) +
  coord_fixed(ratio = 1)
```

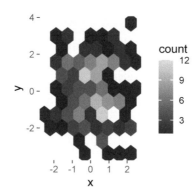

7.5.5 Density functions

Empirical density functions are the equivalent of a histogram, but are continuous and not calculated using bins. They can be estimated in 1 or 2 dimensions (1D or 2D), for x or x and y, respectively. As with histograms it is possible to use different *geometries* to visualize them. Examples of the use of geom_density() to create 1D density plots follow.

```
ggplot(my.data, aes(y, color = group)) +
  geom_density()
```

A semitransparent fill can be used instead of coloured lines.

```
ggplot(my.data, aes(y, fill = group)) +
  geom_density(alpha = 0.5)
```

Examples of 2D density plots follow. In the first example we use two *geometries* which were earlier described, geom_point() and geom_rug(), to plot the observations in the background. With stat_density_2d() we add a two-dimensional density "map" represented using isolines. We map group to the color *aesthetic*.

```
ggplot(my.data, aes(x, y, color = group)) +
  geom_point() +
```

```
geom_rug() +
stat_density_2d()
```

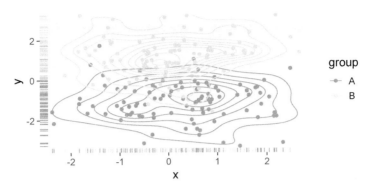

In this case, `geom_density_2d()` is equivalent, and we can replace it in the last line in the chunk above.

```
geom_density_2d()
```

In the next example we plot the groups in separate panels, and use a *geometry* supporting the `fill` *aesthetic* and we map to it the variable `level`, computed by `stat_density_2d()`

```
ggplot(my.data, aes(x, y)) +
stat_density_2d(aes(fill = stat(level)), geom = "polygon") +
  facet_wrap(~group)
```

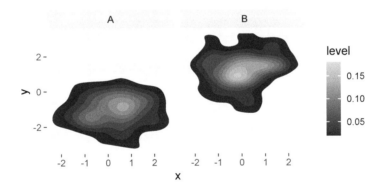

7.5.6 Box and whiskers plots

Box and whiskers plots, also very frequently called just box plots, are also summaries that convey some of the properties of a distribution. They are calculated and plotted by means of `stat_boxplot()` or its matching `geom_boxplot()`. Although they can be calculated and plotted based on just a few observations, they are not useful unless each box plot is based on more than 10 to 15 observations.

```
ggplot(my.data, aes(group, y)) +
  stat_boxplot()
```

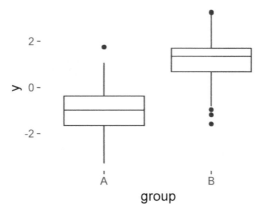

As with other *statistics,* their appearance obeys both the usual *aesthetics* such as `color`, and parameters specific to this type of visual representation: `outlier.color`, `outlier.fill`, `outlier.shape`, `outlier.size`, `outlier.stroke` and `outlier.alpha`, which affect the outliers in a way similar to the equivalent aethetics in `geom_point()`. The shape and width of the "box" can be adjusted with `notch`, `notchwidth` and `varwidth`. Notches in a boxplot serve a similar role for comparing medians as confidence limits serve when comparing means.

```
ggplot(my.data, aes(group, y)) +
  stat_boxplot(notch = TRUE, width = 0.4,
              outlier.color = "red", outlier.shape = "*", outlier.size = 5)
```

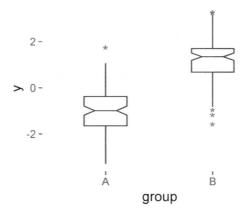

7.5.7   Violin plots

Violin plots are a more recent development than box plots, and usable with relatively large numbers of observations. They could be thought of as being a sort of hybrid between an empirical density function (see section 7.5.5 on page 248) and a box plot (see section 7.5.6 on page 249). As is the case with box plots, they are particularly useful when comparing distributions of related data, side by side. They can be created with `geom_violin()` as shown in the examples below.

```
ggplot(my.data, aes(group, y)) +
  geom_violin()
```

```
ggplot(my.data, aes(group, y, fill = group)) +
  geom_violin(alpha = 0.16) +
  geom_point(alpha = 0.33, size = 1.5,
             color = "black", shape = 21)
```

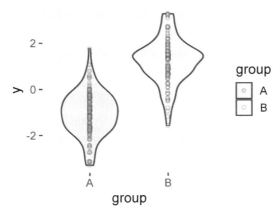

As with other *geometries*, their appearance obeys both the usual *aesthetics* such as color, and others specific to these types of visual representation.

Other types of displays related to violin plots are *beeswarm* plots and *sina* plots, and can be produced with *geometries* defined in packages 'ggbeeswarm' and 'ggforce', respectively. A minimal example of a beeswarm plot is shown below. See the documentation of the packages for details about the many options in their use.

```
ggplot(my.data, aes(group, y)) +
  geom_quasirandom()
```

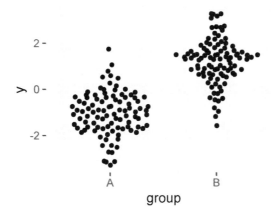

7.6 Facets

Facets are used in a special kind of plots containing multiple panels in which the panels share some properties. These sets of coordinated panels are a useful tool for visualizing complex data. These plots became popular through the `trellis` graphs in S, and the 'lattice' package in R. The basic idea is to have rows and/or columns of plots with common scales, all plots showing values for the same response variable. This is useful when there are multiple classification factors in a data set. Similar-looking plots, but with free scales or with the same scale but a 'floating' intercept, are sometimes also useful. In 'ggplot2' there are two possible types of facets: facets organized in a grid, and facets along a single 'axis' of variation but, possibly, wrapped into two or more rows. These are produced by adding `facet_grid()` or `facet_wrap()`, respectively. In the examples below we use `geom_point()` but faceting can be used with `ggplot` objects containing diverse kinds of layers, displaying either observations or summaries from `data`.

We start by creating and saving a single-panel plot that we will use through this section to demonstrate how the same plot changes when we add facets.

```
p <- ggplot(data = mtcars, aes(wt, mpg)) +
  geom_point()
p
```

A grid of panels has two dimensions, `rows` and `cols`. These dimensions in the grid of plot panels can be "mapped" to factors. Until recently a formula syntax was the only available one. Although this notation has been retained, the preferred syntax is currently to use the parameters `rows` and `cols`. We use `cols` in this example. Note that we need to use `vars()` to enclose the names of the variables in the data. The "headings" of the panels or *strip labels* are by default the levels of the factors.

```
p + facet_grid(cols = vars(cyl))
```

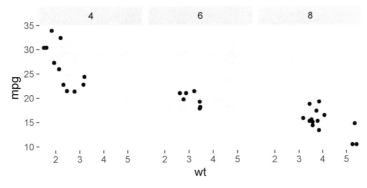

In the "historical notation" the same plot would have been coded as follows.

```
p + facet_grid(. ~ cyl)
```

By default, all panels share the same scale limits and share the plotting space evenly, but these defaults can be overridden.

```
p + facet_grid(cols = vars(cyl), scales = "free")
p + facet_grid(cols = vars(cyl), scales = "free", space = "free")
```

To obtain a 2D grid we need to specify both rows and cols.

```
p + facet_grid(rows = vars(vs), cols = vars(am))
```

Margins display an additional column or row of panels with the combined data.

```
p + facet_grid(cols = vars(cyl), margins = TRUE)
```

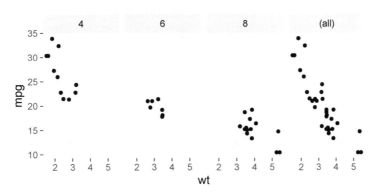

We can represent more than one variable per dimension of the grid of plot panels. For this example, we also override the default labeller used for the panels with one that includes the name of the variable in addition to factor levels in the *strip labels*.

```
p + facet_grid(cols = vars(vs, am), labeller = label_both)
```

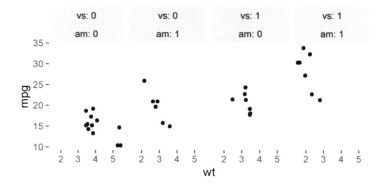

🍵 Sometimes we may want to have mathematical expressions or Greek letters in the panel headings. The next example shows a way of achieving this. The key is to use as `labeller` a function that parses character strings into R expressions.

```
mtcars$cyl12 <- factor(mtcars$cyl,
                       labels = c("alpha", "beta", "sqrt(x, y)"))
p1 <- ggplot(data = mtcars, aes(mpg, wt)) +
      geom_point() +
      facet_grid(cols = vars(cyl12), labeller = label_parsed)
```

More frequently we may need to include the levels of the factor used in the faceting as part of the labels. Here we use as `labeller`, function `label_bquote()` with a special syntax that allows us to use an expression where replacement based on the facet (panel) data takes place. See section 7.12 for an example of the use of `bquote()`, the R function on which `label_bquote()`, is built.

```
p +
  facet_grid(cols = vars(cyl),
             labeller = label_bquote(cols = .(cyl)~"cylinders"))
```

In the next example we create a plot with wrapped facets. In this case the number of levels is small, and no wrapping takes place by default. In cases when more panels are present, wrapping into two or more continuation rows is the default. Here, we force wrapping with `nrow = 2`. When using `facet_wrap()` there is only one dimension, and the parameter is called `facets`, instead of `rows` or `cols`.

```
p + facet_wrap(facets = vars(cyl), nrow = 2)
```

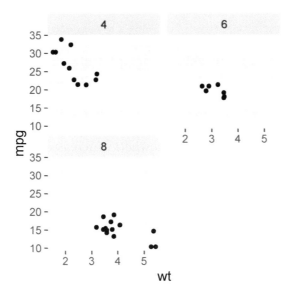

The example below (plot not shown), is similar to the earlier one for facet_grid, but faceting according to two factors with facet_wrap() along a single wrapped row of panels.

```
p + facet_wrap(facets = vars(vs, am), nrow = 2, labeller = label_both)
```

7.7 Scales

In earlier sections of this chapter, examples have used the default *scales* or we have set them with convenience functions. In the present section we describe in more detail the use of *scales*. There are *scales* available for different *aesthetics* (≈ attributes) of the plotted geometrical objects, such as position (x, y, z), size, shape, linetype, color, fill, alpha or transparency, angle. Scales determine how values in data are mapped to values of an *aesthetics*, and how these values are labeled.

Depending on the characteristics of the data being mapped, *scales* can be continuous or discrete, for numeric or factor variables in data, respectively. On the other hand, some *aesthetics*, like size, can vary continuously but others like linetype are inherently discrete. In addition to discrete scales for inherently discrete *aesthetics*, discrete scales are available for those *aesthetics* that are inherently continuous, like x, y, size, color, etc.

The scales used by default set the mapping automatically (e.g., which color value corresponds to $x = 0$ and which one to $x = 1$). However, for each *aesthetic* such as color, there are multiple scales to choose from when creating a plot, both continuous and discrete (e.g., 20 different color scales in 'ggplot2' 3.2.0).

⌨ *Aesthetics* in a plot layer, in addition to being determined by mappings, can also be set to constant values (e.g., plotting all points in a layer in red instead of the default black). *Aesthetics* set to constant values, are not mapped to data, and are consequently independent of scales. In other words, properties of plot elements can be either set to a single constant value of an *aesthetic* affecting all observations present in the layer `data`, or mapped to a variable in `data` in which case the value of the *aesthetic*, such as `color`, will depend on the values of the mapped variable.

The most direct mapping to data is `identity`, which means that the data is taken at its face value. In a color scale, say `scale_color_identity()`, the variable in the data would be encoded with values such as "red", "blue"—i.e., valid R colours. In a simple mapping using `scale_color_discrete()` levels of a factor, such as "treatment" and "control" would be represented as distinct colours with the correspondence of individual factor levels to individual colours selected automatically by default. In contrast with `scale_color_manual()` the user needs to explicitly provide the mapping between factor levels and colours by passing arguments to the scale functions' parameters `breaks` and `values`.

A continuous data variable needs to be mapped to an *aesthetic* through a continuous scale such as `scale_color_continuous()` or one its various variants. Values in a `numeric` variable will be mapped into a continuous range of colours, determined either automatically through a palette or manually by giving the colours at the extremes, and optionally at multiple intermediate values, within the range of variation of the mapped variable (e.g., scale settings so that the color varies gradually between "red" and "gray50"). Handling of missing values is such that mapping a value in a variable to an NA value for an aesthetic such as color makes the mapped values invisible. The reverse, mapping NA values in the data to a specific value of an aesthetic is also possible (e.g., displaying NA values in the mapped variable in red, while other values are mapped to shades of blue).

7.7.1 Axis and key labels

First we describe a feature common to all scales, their `name`. The default `name` of all scales is the name of the variable or the expression mapped to it. In the case of the x, y and z *aesthetics* the `name` given to the scale is used for the axis labels. For other *aesthetics* the name of the scale becomes the "heading" or *key title* of the guide or key. All scales have a `name` parameter to which a character string or R expression (see section 7.12) can be passed as an argument to override the default.

Whole-plot title, subtitle and caption are not connected to *scales* or *data*. A title (`label`) and `subtitle` can be added least confusingly with function `ggtitle()` by passing either character strings or R expressions as arguments.

```
ggplot(data = Orange,
       aes(x = age, y = circumference, color = Tree)) +
  geom_line() +
  geom_point() +
  expand_limits(y = 0) +
```

```
scale_x_continuous(name = "Time (d)") +
scale_y_continuous(name = "Circumference (mm)") +
ggtitle(label = "Growth of orange trees",
        subtitle = "Starting from 1968-12-31")
```

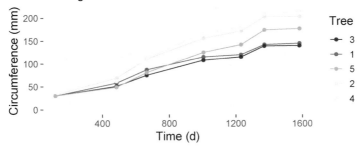

Convenience functions `xlab()` and `ylab()` can be used to set the axis labels to match those in the previous chunk.

```
xlab("Time (d)") +
ylab("Circumference (mm)") +
```

Convenience function `labs()` is useful when we use default scales for all the *aesthetics* in a plot but want to manually set axis labels and/or key titles—i.e., the name of these scales. `labs()` accepts arguments for these names using, as parameter names, the names of the *aesthetics*. It also allows us to set `title`, `subtitle`, `caption` and `tag`, of which the first two can also be set with `ggtitle()`.

```
ggplot(data = Orange,
       aes(x = age, y = circumference, color = Tree)) +
  geom_line() +
  geom_point() +
  expand_limits(y = 0) +
  labs(title = "Growth of orange trees",
       subtitle = "Starting from 1968-12-31",
       caption = "see Draper, N. R. and Smith, H. (1998)",
       tag = "A",
       x = "Time (d)",
       y = "Circumference (mm)",
       color = "Tree\nnumber")
```

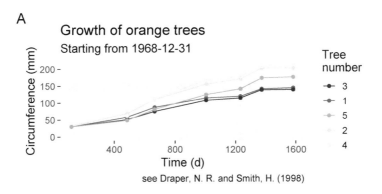

> **ℹ** Make an empty plot (`ggplot()`) and add to it as title an R expression producing $y = b_0 + b_1 x + b_2 x^2$. (Hint: have a look at the examples for the use of expressions in the `plotmath` demo in R by typing `demo(plotmath)` at the R console.

7.7.2 Continuous scales

We start by listing the most frequently used arguments to the continuous scale functions: `name`, `breaks`, `minor_breaks`, `labels`, `limits`, `expand`, `na.value`, `trans`, `guide`, and `position`. The value of `name` is used for axis labels or the key title (see previous section). The arguments to `breaks` and `minor_breaks` override the default locations of major and minor ticks and grid lines. Setting them to `NULL` suppresses the ticks. By default the tick labels are generated from the value of `breaks` but an argument to `labels` of the same length as `breaks` will replace these defaults. The values of `limits` determine both the range of values in the data included and the plotting area as described above—by default the out-of-bounds (oob) observations are replaced by `NA` but it is possible to instead "squish" these observations towards the edge of the plotting area. The argument to `expand` determines the size of the margins or padding added to the area delimited by `lims` when setting the "visual" plotting area. The value passed to `na.value` is used as a replacement for `NA` valued observations—most useful for `color` and `fill` aesthetics. The transformation object passed as an argument to `trans` determines the transformation used—the transformation affects the rendering, but breaks and tick labels remain expressed in the original data units. The argument to `guide` determines the type of key or removes the default key. Depending on the scale in question not all these parameters are available.

We generate new fake data.

```
fake2.data <-
  data.frame(y = c(rnorm(20, mean = 20, sd = 5),
                   rnorm(20, mean = 40, sd = 10)),
             group = factor(c(rep("A", 20), rep("B", 20))),
             z = rnorm(40, mean = 12, sd = 6))
```

7.7.2.1 Limits

Limits are relevant to all kinds of *scales*. Limits are set through parameter `limits` of the different scale functions. They can also be set with convenience functions `xlim()` and `ylim()` in the case of the x and y *aesthetics*, and more generally with function `lims()` which like `labs()`, takes arguments named according to the name of the *aesthetics*. The `limits` argument of scales accepts vectors, factors or a function computing them from `data`. In contrast, the convenience functions do not accept functions as their arguments.

In the next example we set "hard" limits, which will exclude some observations from the plot and from any computation of summaries or fitting of smoothers.

More exactly, the off-limits observations are converted to NA values before they are passed as data to *geometries*.

```
ggplot(fake2.data, aes(z, y)) + geom_point() +
  scale_y_continuous(limits = c(0, 100))
```

To set only one limit leaving the other free, we can use NA as a boundary.

```
scale_y_continuous(limits = c(50, NA))
```

Convenience functions ylim() and xlim() can be used to set the limits to the default *x* and *y* scales in use. We here use ylim(), but xlim() is identical except for the *scale* it affects.

```
ylim(50, NA)
```

In general, setting hard limits should be avoided, even though a warning is issued about NA values being omitted, as it is easy to unwillingly subset the data being plotted. It is preferable to use function expand_limits() as it safely *expands* the dynamically computed default limits of a scale—the scale limits will grow past the requested expanded limits when needed to accommodate all observations. The arguments to x and y are numeric vectors of length one or two each, matching how the limits of the *x* and *y* continuous scales are defined. Here we expand the limits to include the origin.

```
ggplot(fake2.data, aes(z, y)) +
  geom_point() +
  expand_limits(y = 0, x = 0)
```

The expand parameter of the scales plays a different role than expand_limits(). It controls how much larger the "visual" plotting area is compared to the limits of the actual plotting area. In other words, it adds a "margin" or padding to the plotting area outside the limits set either dynamically or manually. Very rarely plots are drawn so that observations are plotted on top of the axes, avoiding this is a key role of expand. Rug plots and marginal annotations will also require the plotting area to be expanded. In 'ggplot2' the default is to always apply some expansion.

We here set the upper limit of the plotting area to be expanded by adding padding to the top and remove the default padding from the bottom of the plotting area.

```
ggplot(fake2.data,
  aes(fill = group, color = group, x = y)) +
  stat_density(alpha = 0.3) +
  scale_y_continuous(expand = expand_scale(add = c(0, 0.02)))
```

Here we instead use a multiplier to a similar effect as above; we add 10% compared to the range of the `limits`.

```
scale_y_continuous(expand = expand_scale(mult = c(0, 0.1)))
```

In the case of scales, we cannot reverse their direction through the setting of limits. We need instead to use a transformation as described in section 7.7.2.3 on page 261. But, inconsistently, `xlim()` and `ylim()` do implicitly allow this transformation through the numeric values passed as limits.

Test what the result is when the first limit is larger than the second one. Is it the same as when setting these same values as limits with `ylim()`?

```
ggplot(fake2.data, aes(z, y)) + geom_point() +
  scale_y_continuous(limits = c(100, 0))
```

7.7.2.2   Ticks and their labels

Parameter `breaks` is used to set the location of ticks along the axis. Parameter `labels` is used to set the tick labels. Both parameters can be passed either a vector or a function as an argument. The default is to compute "good" breaks based on the limits and format the numbers as strings.

When manually setting breaks, we can keep the default computed labels for the breaks.

```
ggplot(fake2.data, aes(z, y)) +
  geom_point() +
  scale_y_continuous(breaks = c(20, pi * 10, 40, 60))
```

The default breaks are computed by function `pretty_breaks()` from 'scales'. The argument passed to its parameter n determines the target number ticks to be generated automatically, but the actual number of ticks computed may be slightly different depending on the range of the data.

```
scale_y_continuous(breaks = scales::pretty_breaks(n = 7))
```

We can set tick labels manually, in parallel to the setting of `breaks` by passing as arguments two vectors of equal length. In the next example we use an expression to obtain a Greek letter.

```
ggplot(fake2.data, aes(z, y)) +
  geom_point() +
  scale_y_continuous(breaks = c(20, pi * 10, 40, 60),
                     labels = c("20", expression(10*pi), "40", "60"))
```

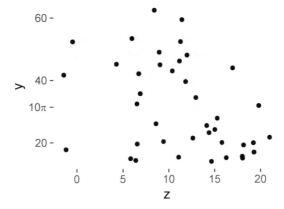

Package 'scales' provides several functions for the automatic generation of labels. For example, to display tick labels as percentages for data available as decimal fractions, we can use function scales::percent().

```
ggplot(fake2.data, aes(z, y / max(y))) +
  geom_point() +
  scale_y_continuous(labels = scales::percent)
```

For currency, we can use scales::dollar(), to include commas separating thousands, millions, so on, we can use scales::comma(), and for numbers formatted using exponents of 10—useful for logarithmic-transformed scales—we can use scales::scientific_format(). It is also possible to use user-defined functions both for breaks and labels.

7.7.2.3 Transformed scales

The default scales used by the x and y aesthetics, scale_x_continuous() and scale_y_continuous(), accept a user-supplied transformation function as an argument to trans with default codetrans = "identity" (no transformation). In addition, there are predefined convenience scale functions for log10, sqrt and reverse.

 Similar to the maths functions of R, the name of the scales are `scale_x_log10()` and `scale_y_log10()` rather than `scale_y_log()` because in R, the function `log` returns the natural logarithm.

We can use `scale_x_reverse()` to reverse the direction of a continuous scale,

```
ggplot(fake2.data, aes(z, y)) +
  geom_point() +
  scale_x_reverse()
```

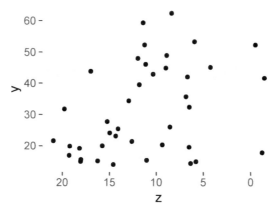

Axis tick-labels display the original values before applying the transformation. The "`breaks`" need to be given in the original scale as well. We use `scale_y_log10()` to apply a $\log_{10}$ transformation to the y values.

```
scale_y_log10(breaks=c(10,20,50,100))
```

Using a transformation in a scale is not equivalent to applying the same transformation on the fly when mapping a variable to the x (or y) *aesthetic* as this results in tick-labels expressed in transformed values.

```
ggplot(fake2.data, aes(z, log10(y))) +
  geom_point()
```

We show next how to specify a transformation to a continuous scale, using a predefined "transformation" object.

```
scale_y_continuous(trans = "reciprocal")
```

Natural logarithms are important in growth analysis as the slope against time gives the relative growth rate. We show this with the `Orange` data set.

```
ggplot(data = Orange,
       aes(x = age, y = circumference, color = Tree)) +
  geom_line() +
  geom_point() +
  scale_y_continuous(trans = "log", breaks = c(20, 50, 100, 200))
```

7.7.2.4   Position of x and y axes

The default position of axes can be changed through parameter `position`, using character constants `"bottom"`, `"top"`, `"left"` and `"right"`.

```
ggplot(data = mtcars, aes(wt, mpg)) +
  geom_point() +
  scale_x_continuous(position = "top") +
  scale_y_continuous(position = "right")
```

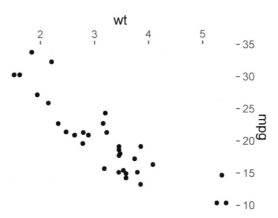

7.7.2.5   Secondary axes

It is also possible to add secondary axes with ticks displayed in a transformed scale.

```
ggplot(data = mtcars, aes(wt, mpg)) +
  geom_point() +
  scale_y_continuous(sec.axis = sec_axis(~ . ^-1, name = "1/y") )
```

It is also possible to use different `breaks` and `labels` than for the main axes, and to provide a different `name` to be used as a secondary axis label.

```
scale_y_continuous(sec.axis = sec_axis(~ . / 2.3521458, name = expression(km / l),
                                       breaks = c(5, 7.5, 10, 12.5)))
```

7.7.3   Time and date scales for x and y

In R and many other computing languages, time values are stored as integer or numeric values subject to special interpretation. Times stored as objects of class POSIXct can be mapped to continuous *aesthetics* such as x and y. Special scales are available for these quantities.

We can set limits and breaks using constants as time or dates. These are most easily input with the functions in packages 'lubridate' or 'anytime'.

⚠️ Warnings are issued in the next two chunks as we are using scale limits to subset a part of the observations present in `data`.

```
ggplot(data = weather_wk_25_2019.tb,
       aes(with_tz(time, tzone = "EET"), air_temp_C)) +
  geom_line() +
  scale_x_datetime(name = NULL,
                   breaks = ymd_hm("2019-06-11 12:00", tz = "EET") + days(0:1),
                   limits = ymd_hm("2019-06-11 00:00", tz = "EET") + days(c(0, 2))) +
  scale_y_continuous(name = "Air temperature (C)") +
  expand_limits(y = 0)
```

```
## Warning: Removed 7199 row(s) containing missing values (geom_path).
```

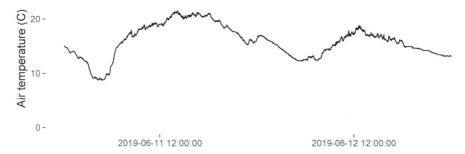

By default the tick labels produced and their formatting are automatically selected based on the extent of the time data. For example, if we have all data collected within a single day, then the tick labels will show hours and minutes. If we plot data for several years, the labels will show the date portion of the time instant. The default is frequently good enough, but it is possible, as for numbers, to use different formatter functions to generate the tick labels.

```
ggplot(data = weather_wk_25_2019.tb,
       aes(with_tz(time, tzone = "EET"), air_temp_C)) +
  geom_line() +
  scale_x_datetime(name = NULL,
                   date_breaks = "1 hour",
                   limits = ymd_hm("2019-06-16 00:00", tz = "EET") + hours(c(6, 18)),
                   date_labels = "%H:%M") +
  scale_y_continuous(name = "Air temperature (C)") +
  expand_limits(y = 0)
```

```
## Warning: Removed 9359 row(s) containing missing values (geom_path).
```

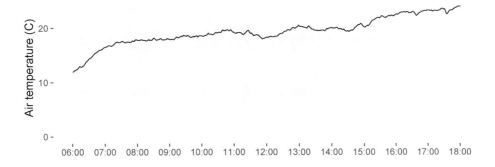

The formatting strings used are those supported by `strptime()` and `help(strptime)` lists them. Change, in the two examples above, the *y*-axis labels used and the limits—e.g., include a single hour or a whole week of data, check which tick labels are produced by default and then pass as an argument to `date_labels` different format strings, taking into account that in addition to the *conversion specification* codes, format strings can include additional text.

7.7.4   Discrete scales for *x* and *y*

In the case of ordered or unordered factors, the tick labels are by default the names of the factor levels. Consequently, one roundabout way of obtaining the desired tick labels is to set them as factor levels. This approach is not recommended as in many cases the text of the desired tick labels may not be recognized as a valid name making the code using them more difficult to type in scripts or at the command prompt. It is best to use simple mnemonic short names for factor levels and variables, and to set suitable labels through *scales* when plotting, as we will show here.

We can use `scale_x_discrete()` to reorder and select the columns without altering the data. If we use this approach to subset the data, then to avoid warnings we need to add `na.rm = TRUE`. We additionally use `scale_x_discrete` to convert level names to uppercase.

```
ggplot(mpg, aes(class, hwy)) +
  stat_summary(geom = "col", fun = mean, na.rm = TRUE) +
  scale_x_discrete(limits = c("compact", "subcompact", "midsize"),
                   labels = c("COMPACT", "SUBCOMPACT", "MIDSIZE"))
```

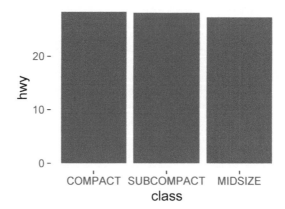

If, as in the previous example, only the case of character strings needs to be changed, passing function `toupper()` or `tolower()` allows a more general and less error-prone approach. In fact any function, user defined or not, which converts the values of `limits` into the desired values can be passed as an argument to `labels`.

```
scale_x_discrete(limits = c("compact", "subcompact", "midsize"),
                 labels = toupper)
```

Alternatively, we can change the order of the columns in the plot by reordering the levels of factor `mpg$class`. This approach makes sense if the ordering needs to be done programmatically based on values in `data`. See section 2.12 on page 56 for details. The example below shows how to reorder the columns, corresponding to the levels of `class` based on the `mean()` of `hwy`.

```
ggplot(mpg, aes(reorder(x = factor(class), X = hwy, FUN = mean), hwy)) +
  stat_summary(geom = "col", fun = mean)
```

7.7.5 Size

For the `size` *aesthetic*, several scales are available, both discrete and continuous. They do not differ much from those already described above. *Geometries* `geom_point()`, `geom_line()`, `geom_hline()`, `geom_vline()`, `geom_text()`, `geom_label()` obey `size` as expected. In the case of `geom_bar()`, `geom_col()`, `geom_area()` and all other geometric elements bordered by lines, `size` is obeyed by these border lines. In fact, other aesthetics natural for lines such as `linetype` also apply to these borders.

When using `size` scales, `breaks` and `labels` affect the key or `guide`. In scales that produce a key passing `guide = FALSE` removes the key corresponding to the scale.

7.7.6 Color and fill

`color` and `fill` scales are similar, but they affect different elements of the plot. All visual elements in a plot obey the `color` *aesthetic*, but only elements that have an inner region and a boundary, obey both `color` and `fill` *aesthetics*. There are separate but equivalent sets of scales available for these two *aesthetics*. We will describe in more detail the `color` *aesthetic* and give only some examples for `fill`. We will, however, start by reviewing how colors are defined and used in R.

7.7.6.1 Color definitions in R

Colors can be specified in R not only through character strings with the names of previously defined colors, but also directly as strings describing the RGB (red, green and blue) components as hexadecimal numbers (on base 16 expressed using 0, 1, 2, 3, 4, 6, 7, 8, 9, A, B, C, D, E, and F as "digits") such as "#FFFFFF" for white or "#000000" for black, or "#FF0000" for the brightest available pure red.

The list of color names known to R can be obtained be typing `colors()` at the R console. Given the number of colors available, we may want to subset them based on their names. Function `colors()` returns a character vector. We can use `grep()` to find the names containing a given character substring, in this example "dark".

```
length(colors())
## [1] 657
```

```
grep("dark",colors(), value = TRUE)
##  [1] "darkblue"        "darkcyan"        "darkgoldenrod"   "darkgoldenrod1"
##  [5] "darkgoldenrod2"  "darkgoldenrod3"  "darkgoldenrod4"  "darkgray"
##  [9] "darkgreen"       "darkgrey"        "darkkhaki"       "darkmagenta"
## [13] "darkolivegreen"  "darkolivegreen1" "darkolivegreen2" "darkolivegreen3"
## [17] "darkolivegreen4" "darkorange"      "darkorange1"     "darkorange2"
## [21] "darkorange3"     "darkorange4"     "darkorchid"      "darkorchid1"
## [25] "darkorchid2"     "darkorchid3"     "darkorchid4"     "darkred"
## [29] "darksalmon"      "darkseagreen"    "darkseagreen1"   "darkseagreen2"
## [33] "darkseagreen3"   "darkseagreen4"   "darkslateblue"   "darkslategray"
## [37] "darkslategray1"  "darkslategray2"  "darkslategray3"  "darkslategray4"
## [41] "darkslategrey"   "darkturquoise"   "darkviolet"
```

To retrieve the RGB values for a color definition we use:

```
col2rgb("purple")
##       [,1]
## red    160
## green   32
## blue   240
```

```
col2rgb("#FF0000")
##       [,1]
## red    255
## green    0
## blue     0
```

Color definitions in R can contain a *transparency* described by an `alpha` value, which by default is not returned.

```
col2rgb("purple", alpha = TRUE)
##       [,1]
## red    160
## green   32
## blue   240
## alpha  255
```

With function `rgb()` we can define new colors. Enter `help(rgb)` for more details.

```
rgb(1, 1, 0)
## [1] "#FFFF00"

rgb(1, 1, 0, names = "my.color")
##  my.color
## "#FFFF00"

rgb(255, 255, 0, names = "my.color", maxColorValue = 255)
##  my.color
## "#FFFF00"
```

As described above, colors can be defined in the RGB *color space*, however, other color models such as HSV (hue, saturation, value) can be also used to define colours.

```
hsv(c(0,0.25,0.5,0.75,1), 0.5, 0.5)
## [1] "#804040" "#608040" "#408080" "#604080" "#804040"
```

Probably a more useful flavor of HSV colors for use in scales are those returned by function `hcl()` for hue, chroma and luminance. While the "value" and "saturation" in HSV are based on physical values, the "chroma" and "luminance" values in HCL are based on human visual perception. Colours with equal luminance will be seen as equally bright by an "average" human. In a scale based on different hues but equal chroma and luminance values, as used by package 'ggplot2', all colours are perceived as equally bright. The hues need to be expressed as angles in degrees, with values between zero and 360.

```
hcl(c(0,0.25,0.5,0.75,1) * 360)
## [1] "#FFC5D0" "#D4D8A7" "#99E2D8" "#D5D0FC" "#FFC5D0"
```

It is also important to remember that humans can only distinguish a limited set of colours, and even smaller color gamuts can be reproduced by screens and printers. Furthermore, variation from individual to individual exists in color perception, including different types of color blindness. It is important to take this into account when choosing the colors used in illustrations.

7.7.7 Continuous color-related scales

Continuous color scales `scale_color_continuous()`, `scale_color_gradient()`, `scale_color_gradient2()`, `scale_color_gradientn()`, `scale_color_date()` and `scale_color_datetime()`, give a smooth continuous gradient between two or more colours. They are used with `numeric`, `date` and `datetime` data. A corresponding set of `fill` scales is also available. Other scales like `scale_color_viridis_c()` and `scale_color_distiller()` are based on the use of ready-made palettes of sets of color gradients chosen to work well together under multiple conditions or for human vision including different types of color blindness.

7.7.8 Discrete color-related scales

Color scales `scale_color_discrete()`, `scale_color_hue()`, `scale_color_gray()` are used with categorical data stored as factors. Other scales like `scale_color_viridis_d()` and `scale_color_brewer()` provide discrete sets of colours based on palettes.

7.7.9 Identity scales

In the case of identity scales, the mapping is one to-one to the data. For example, if we map the `color` or `fill` *aesthetic* to a variable using `scale_color_identity()` or `scale_fill_identity()`, the mapped variable must already contain valid color definitions. In the case of mapping `alpha`, the variable must contain numeric values in the range 0 to 1.

We create a data frame containing a variable `colors` containing character strings interpretable as the names of color definitions known to R. We then use them directly in the plot.

```
df99 <- data.frame(x = 1:10, y = dnorm(10), colors = rep(c("red", "blue"), 5))

ggplot(df99, aes(x, y, color = colors)) +
  geom_point() +
  scale_color_identity()
```

> How does the plot look, if the identity scale is deleted from the example above? Edit and re-run the example code.
>
> While using the identity scale, how would you need to change the code example above, to produce a plot with green and purple points?

7.8 Adding annotations

The idea of annotations is that they add plot elements that are not directly connected with `data`, which we could call "decorations" such as arrows used to highlight some feature of the data, specific points along an axis, etc. They are referenced to the "natural" coordinates used to plot the observations, but are elements that do not represent observations or summaries computed from the observations. Annotations are added to a ggplot with `annotate()` as plot layers (each call to `annotate()` creates a new layer). To achieve the behavior expected of annotations, `annotate()` does not inherit the default `data` or `mapping` of variables to *aesthetics*. Annotations frequently make use of `"text"` or `"label"` *geometries* with character strings as data, possibly to be parsed as expressions. However, for example, the `"segment"` geometry can be used to add arrows.

> ⚠ While layers added to a plot directly using *geometries* and *statistics* respect faceting, annotation layers added with `annotate()` are replicated unchanged in every panel of a faceted plot. The reason is that annotation layers accept *aesthetics* only as constant values which are the same for every panel as no grouping is possible without a `mapping` to `data`.

We show a simple example using `"text"` as *geometry*.

```
ggplot(fake2.data, aes(z, y)) +
  geom_point() +
  annotate(geom = "text",
           label = "origin",
           x = 0, y = 0,
           color = "blue",
           size=4)
```

> 🎚 Play with the values of the arguments to `annotate()` to vary the position, size, color, font family, font face, rotation angle and justification of the annotation.

It is relatively common to use inset tables, plots, bitmaps or vector plots as annotations. With `annotation_custom()`, grobs ('grid' graphical object) can be added to a ggplot. To add another or the same plot as an inset, we first need to convert it into a grob. In the case of a ggplot we use `ggplotGrob()`. In this example the inset is a zoomed-in window into the main plot. In addition to the grob, we need to provide the coordinates expressed in "natural" data units of the main plot for the location of the grob.

```
p <- ggplot(fake2.data, aes(z, y)) +
  geom_point()
p + expand_limits(x = 40) +
  annotation_custom(ggplotGrob(p + coord_cartesian(xlim = c(5, 10), ylim = c(20, 40)) +
                              theme_bw(10)),
                    xmin = 21, xmax = 40, ymin = 30, ymax = 60)
```

This approach has the limitation that if used together with faceting, the inset will be the same for each plot panel. See section 7.4.8 on page 233 for *geometries* that can be used to add insets.

In the next example, in addition to adding expressions as annotations, we also pass expressions as tick labels through the scale. Do notice that we use recycling for setting the breaks, as `c(0, 0.5, 1, 1.5, 2) * pi` is equivalent to `c(0, 0.5 * pi, pi, 1.5 * pi, 2 * pi)`. Annotations are plotted at their own position, unrelated to any observation in the data, but using the same coordinates and units as for plotting the data.

```
ggplot(data.frame(x = c(0, 2 * pi)), aes(x = x)) +
  stat_function(fun = sin) +
  scale_x_continuous(
    breaks = c(0, 0.5, 1, 1.5, 2) * pi,
    labels = c("0", expression(0.5~pi), expression(pi),
               expression(1.5~pi), expression(2~pi))) +
  labs(y = "sin(x)") +
  annotate(geom = "text",
           label = c("+", "-"),
           x = c(0.5, 1.5) * pi, y = c(0.5, -0.5),
           size = 20) +
  annotate(geom = "point",
           color = "red",
           shape = 21,
           fill = "white",
           x = c(0, 1, 2) * pi, y = 0,
           size = 6)
```

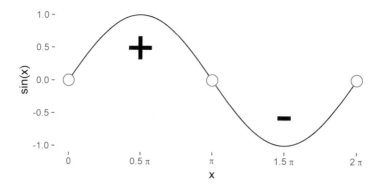

Modify the plot above to show the cosine instead of the sine function, replacing sin with cos. This is easy, but the catch is that you will need to relocate the annotations.

We cannot use annotate() with geom = "vline" or geom = "hline" as we can use geom = "line" or geom = "segment". Instead, geom_vline() and/or geom_hline() can be used directly passing constant arguments to them. See section 7.4.3 on page 224.

7.9   Coordinates and circular plots

Circular plots can be thought of as plots equivalent to those described earlier in this chapter but drawn using a different system of coordinates. This is a key insight, that the grammar of graphics as implemented in 'ggplot2' makes use of. To obtain circular plots we use the same *geometries, statistics* and *scales* we have been using with the default system of cartesian coordinates. The only thing that we need to do is to add coord_polar() to override the default. Of course only some observed quantities can be better perceived in circular plots than in cartesian plots. Here we add a new "word" to the grammar of graphics, *coordinates*, such as coord_polar(). When using polar coordinates, the x and y *aesthetics* correspond to the angle and radial distance, respectively.

7.9.1   Wind-rose plots

Some types of data are more naturally expressed on polar coordinates than on cartesian coordinates. The clearest example is wind direction, from which the name derives. In some cases of time series data with a strong periodic variation, polar coordinates can be used to highlight any phase shifts or changes in frequency. A more mundane application is to plot variation in a response variable through the day with a clock-face-like representation of time of day.

Wind rose plots are frequently histograms or density plots drawn on a polar system of coordinates (see sections 7.5.4 and 7.5.5 on pages 245 and 248, respectively for a description of the use of these *statistics* and *geometries*). We will use them for examples where we plot wind speed and direction data, measured once per minute during 24 h (from package 'learnrbook').

Here we plot a circular histogram of wind directions with 30-degree-wide bins. We use stat_bin(). The counts represent the number of minutes during 24 h when the wind direction was within each bin.

```
p <- ggplot(viikki_d29.dat, aes(WindDir_D1_WVT))  +
  coord_polar() +
  scale_x_continuous(breaks = c(0,  90,  180,  270),
                     labels = c("N",  "E",  "S",  "W"),
                     limits = c(0,  360),
                     expand = c(0,  0),
                     name = "Wind direction")
p + stat_bin(color = "black",  fill = "gray50",  geom = "bar",
             binwidth = 30,  na.rm = TRUE) + labs(y = "Frequency")
```

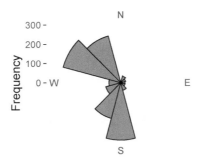

Wind direction

For an equivalent plot, using an empirical density, we have to use stat_density() instead of stat_bin(), geom_polygon() instead of geom_bar() and change the name of the y scale.

```
p + stat_density(color = "black",  fill = "gray50",
                 geom = "polygon",  size = 1) + labs(y = "Density")
```

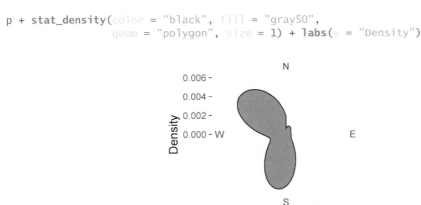

Wind direction

As the final wind-rose plot example, we do 2D density plot with facets added with `facet_wrap()` to have separate panels for AM and PM. This plot uses fill to describe the density of observations for different combinations wind directions and speeds, the radius (*y aesthetic*) to represent wind speeds and the angle (*x aesthetic*) to represent wind direction.

```
ggplot(viikki_d29.dat, aes(WindDir_D1_WVT, WindSpd_S_WVT)) +
  coord_polar() +
  stat_density_2d(aes(fill = stat(level)), geom = "polygon") +
  scale_x_continuous(breaks = c(0, 90, 180, 270),
                     labels = c("N", "E", "S", "W"),
                     limits = c(0, 360),
                     expand = c(0, 0),
                     name = "Wind direction") +
  scale_y_continuous(name = "Wind speed (m/s)") +
  facet_wrap(~factor(ifelse(hour(solar_time) < 12, "AM", "PM")))
```

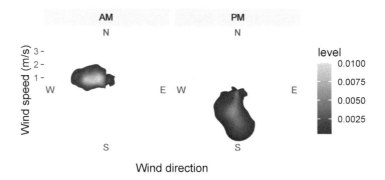

7.9.2 Pie charts

> ⚠️ Pie charts are more difficult to read than bar charts because our brain is better at comparing lengths than angles. If used, pie charts should only be used to show composition, or fractional components that add up to a total. In this case, used only if the number of "pie slices" is small (rule of thumb: seven at most), however in general, they are best avoided.

As we use `geom_bar()` which defaults to use `stat_count`. We use the brewer scale for nice colors.

```
ggplot(data = mpg, aes(x = factor(1), fill = factor(class))) +
  geom_bar(width = 1, color = "black") +
  coord_polar(theta = "y") +
  scale_fill_brewer() +
  scale_x_discrete(breaks = NULL) +
  labs(x = NULL, fill = "Vehicle class")
```

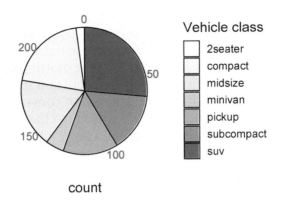

Edit the code for the pie chart above to obtain a bar chart. Which one of the two plots is easier to read?

7.10   Themes

In 'ggplot2', *themes* are the equivalent of style sheets. They determine how the different elements of a plot are rendered when displayed, printed or saved to a file. *Themes* do not alter what aesthetics or scales are used to plot the observations or summaries, but instead how text-labels, titles, axes, grids, plotting-area background and grid, etc., are formatted and if displayed or not. Package 'ggplot2' includes several predefined *theme constructors* (usually described as *themes*), and independently developed extension packages define additional ones. These constructors return complete themes, which when added to a plot, replace any theme already present in whole. In addition to choosing among these already available *complete themes*, users can modify the ones already present by adding *incomplete themes* to a plot. When used in this way, *incomplete themes* usually are created on the fly. It is also possible to create new theme constructors that return complete themes, similar to theme_gray() from 'ggplot2'.

7.10.1   Complete themes

The theme used by default is theme_gray() with default arguments. In 'ggplot2', predefined themes are defined as constructor functions, with parameters. These parameters allow changing some "base" properties. The base_size for text elements controlled is given in points, and affects all text elements in the returned theme object as the size of these elements is by default defined relative to the base size. Another parameter, base_family, allows the font family to be set. These functions return complete themes.

⚠ *Themes* have no effect on layers produced by *geometries* as themes have no effect on *mappings*, *scales* or *aesthetics*. In the name `theme_bw()` black-and-white refers to the color of the background of the plotting area and labels. If the *color* or fill *aesthetics* are mapped or set to a constant in the figure, these will be respected irrespective of the theme. We cannot convert a color figure into a black-and-white one by adding a *theme*, we need to change the *aesthetics* used, for example, use `shape` instead of `color` for a layer added with `geom_point()`.

Even the default `theme_gray()` can be added to a plot, to modify it, if arguments different to the defaults are passed when called. In this example we override the default base size with a larger one and the default sans-serif font with one with serifs.

```
ggplot(fake2.data, aes(z, y)) +
  geom_point() +
  theme_gray(base_size = 15,
             base_family = "serif")
```

🎚 Change the code in the previous chunk to use, one at a time, each of the predefined themes from 'ggplot2': `theme_bw()`, `theme_classic()`, `theme_minimal()`, `theme_linedraw()`, `theme_light()`, `theme_dark()` and `theme_void()`.

☕ Predefined "themes" like `theme_gray()` are, in reality, not themes but instead are constructors of theme objects. The *themes* they return when called depend on the arguments passed to their parameters. In other words, `theme_gray(base_size = 15)`, creates a different theme than `theme_gray(base_size = 11)`. In this case, as sizes of different text elements are defined relative to the base size, the size of all text elements changes in co-ordination. Font size changes by *themes* do not affect the size of text or labels in plot layers created with geometries, as their size is controlled by the `size` *aesthetic*.

A frequent idiom is to create a plot without specifying a theme, and then adding the theme when printing or saving it. This can save work, for example, when producing different versions of the same plot for a publication and a talk.

```
p <- ggplot(fake2.data, aes(z, y)) +
     geom_point()
print(p + theme_bw())
```

It is also possible to change the theme used by default in the current R session with `theme_set()`.

```
old_theme <- theme_set(theme_bw(15))
```

Similar to other functions used to change options in R, `theme_set()` returns the previous setting. By saving this value to a variable, here `old_theme`, we are able to restore the previous default, or undo the change.

```
theme_set(old_theme)
p
```

7.10.2   Incomplete themes

If we want to extensively modify a theme, and/or reuse it in multiple plots, it is best to create a new constructor, or a modified complete theme as described in the next section. In other cases we may need to tweak some theme settings for a single figure, in which case we can most effectively do this when creating a plot. We exemplify this approach by solving the problem of overlapping x-axis tick labels. In practice this problem is most frequent when factor levels have long names or the labels are dates. Rotating the tick labels is the most elegant solution from the graphics design point of view.

```
ggplot(fake2.data, aes(z + 1000, y)) +
  geom_point() +
  scale_x_continuous(breaks = scales::pretty_breaks(n = 8)) +
  theme(axis.text.x = element_text(angle = 90, hjust = 1, vjust = 0.5))
```

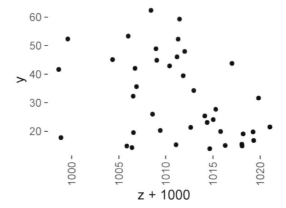

When tick labels are rotated, one usually needs to set both the horizontal and vertical justification, `hjust` and `vjust`, as the default values stop being suitable. This is due to the fact that justification settings are referenced to the text itself rather than to the plot, i.e., **vertical** justification of x-axis tick labels rotated 90 degrees shifts their alignment with respect to tick marks along the (**horizontal**) x axis.

Play with the code in the last chunk above, modifying the values used for `angle`, `hjust` and `vjust`. (Angles are expressed in degrees, and justification with values between 0 and 1).

A less elegant approach is to use a smaller font size. Within `theme()`, function `rel()` can be used to set size relative to the base size. In this example, we use `axis.text.x` so as to change the size of tick labels only for the x axis.

```
theme(axis.text.x = element_text(size = rel(0.6)))
```

Theme definitions follow a hierarchy, allowing us to modify the formatting of groups of similar elements, as well as of individual elements. In the chunk above, had we used `axis.text` instead of `axis.text.x`, the change would have affected the tick labels in both x and y axes.

Modify the example above, so that the tick labels on the x-axis are blue and those on the y-axis red, and the font size is the same for both axes, but changed from the default. Consult the documentation for `theme()` to find out the names of the elements that need to be given new values. For examples, see *ggplot2: Elegant Graphics for Data Analysis* (Wickham and Sievert 2016) and *R Graphics Cookbook* (Chang 2018).

Formatting of other text elements can be adjusted in a similar way, as well as thickness of axes, length of tick marks, grid lines, etc. However, in most cases these are graphic design elements that are best kept consistent throughout sets of plots and best handled by creating a new *theme* that can be easily reused.

> ⚠ If you both add a *complete theme* and want to modify some of its elements, you should add the whole theme before modifying it with + `theme(...)`. This may seem obvious once one has a good grasp of the grammar of graphics, but can be at first disconcerting.

It is also possible to modify the default theme used for rendering all subsequent plots.

```
old_theme <- theme_update(text = element_text(color = "darkred"))
```

7.10.3 Defining a new theme

Themes can be defined both from scratch, or by modifying existing saved themes, and saving the modified version. As discussed above, it is also possible to define a new, parameterized theme constructor function.

Unless we plan to widely reuse the new theme, there is usually no need to define a new function. We can simply save the modified theme to a variable and add it to different plots as needed. As we will be adding a "ready-build" theme object rather than a function, we do not use parentheses.

```
my_theme <- theme_bw() + theme(text = element_text(color = "darkred"))
p + my_theme
```

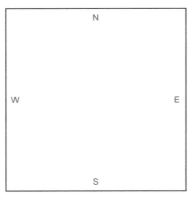

Wind direction

> 📖 It is always good to learn to recognize error messages. One way of doing this is by generating errors on purpose. So do add parentheses to the statement in the code chunk above and study the error message.

How to create a new theme constructor similar to those in package 'ggplot2' can be fairly simple if the changes are few. As the implementation details of theme objects may change in future versions of 'ggplot2', the safest approach is to rely only on the public interface of the package. We can "wrap" the functions exported by package 'ggplot2' inside a new function. For this we need to find out what are the parameters and their order and duplicate these in our wrapper. Looking at the "usage" section of the help page for `theme_gray()` is enough. In this case, we retain compatibility, but add a new base parameter, `base_color`, and set a different default for `base_family`. The key detail is passing `complete = TRUE` to `theme()`, as this tags the returned theme as being usable by itself, resulting in replacement of any theme already in a plot when it is added.

```
my_theme_gray <-
  function (base_size = 11,
            base_family = "serif",
            base_line_size = base_size/22,
            base_rect_size = base_size/22,
            base_color = "darkblue") {
    theme_gray(base_size = base_size,
               base_family = base_family,
               base_line_size = base_line_size,
               base_rect_size = base_rect_size) +
    theme(line = element_line(color = base_color),
          rect = element_rect(color = base_color),
          text = element_text(color = base_color),
          title = element_text(color = base_color),
          axis.text = element_text(color = base_color), complete = TRUE)
  }
```

In the chunk above we have created our own theme constructor, without too much effort, and using an approach that is very likely to continue working with future versions of 'ggplot2'. The saved theme is a function with parameters and defaults for them. In this example we have kept the function parameters the same as those used in 'ggplot2', only adding an additional parameter after the existing ones to maximize compatibility and avoid surprising users. To avoid surprising users, we may want additionally to make `my_theme_gray()` a synonym of `my_theme_gray()` following 'ggplot2' practice.

```
my_theme_gray <- my_theme_gray
```

Finally, we use the new theme constructor in the same way as those defined in 'ggplot2'.

```
p + my_theme_gray(15, base_color = "darkred")
```

```
                              N



           W                              E



                              S
                       Wind direction
```

7.11 Composing plots

In section 7.6 on page 252, we described how facets can be used to created coordinated sets of panels, based on a single data set. Rather frequently, we need to assemble a composite plot from individually created plots. If one wishes to have correctly aligned axis labels and plotting areas, similar to when using facets, then the task is not easy to achieve without the help of especial tools.

Package 'patchwork' defines a simple grammar for composing plots created with 'ggplot2'. We briefly describe here the use of operators +, | and /, although 'patchwork' provides additional tools for defining complex layouts of panels. While + allows different layouts, | composes panels side by side, and / composes panels on top of each other. The plots to be used as panels can be grouped using parentheses.

We start by creating and saving three plots.

```
p1 <- ggplot(mpg, aes(displ, cty, color = factor(cyl))) +
        geom_point() +
        theme(legend.position = "top")
p2 <- ggplot(mpg, aes(displ, cty, color = factor(year))) +
        geom_point() +
        theme(legend.position = "top")
p3 <- ggplot(mpg, aes(factor(model), cty)) +
        geom_point() +
        theme(axis.text.x =
                element_text(angle = 90, hjust = 1, vjust = 0.5))
```

Next, we compose a plot using as panels the three plots created above (plot not shown).

```
(p1 | p2) / p3
```

We add a title and tag the panels with a letter. In this, and similar cases, parentheses may be needed to alter the default precedence of the R operators.

```
((p1 | p2) / p3) +
    plot_annotation(title = "Fuel use in city traffic:", tag_levels = 'a')
```

Fuel use in city traffic:

7.12 Using `plotmath` expressions

In sections 7.5.1 and 7.4.7 we gave some simple examples of the use of R expressions in plots. The `plotmath` demo and help in R provide enough information to start using expressions in plots. However, composing syntactically correct expressions can be challenging because their syntax is rather unusual. Although expressions are shown here in the context of plotting, they are also used in other contexts in R code.

In general it is possible to create *expressions* explicitly with function expression(), or by parsing a character string. In the case of 'ggplot2' for some plot elements, layers created with geom_text and geom_label, and the strip labels of facets the parsing is delayed and applied to mapped character variables in data. In contrast, for titles, subtitles, captions, axis-labels, etc. (anything that is defined within labs()) the expressions have to be entered explicitly, or saved as such into a variable, and the variable passed as an argument.

When plotting expressions using geom_text(), that character strings are to be parsed is signaled with parse = TRUE. In the case of facets' strip labels, parsing or not depends on the *labeller* function used. An additional twist is in this case the possibility of combining static character strings with values taken from data.

The most difficult thing to remember when writing expressions is how to connect the different parts. A tilde (~) adds space in between symbols. Asterisk (*) can be also used as a connector, and is needed usually when dealing with numbers. Using space is allowed in some situations, but not in others. To include bits of text within an expression we need to use quotation marks. For a long list of examples have a look at the output and code displayed by demo(plotmath) at the R command prompt.

We will use a couple of complex examples to show how to use expressions for different elements of a plot. We first create a data frame, using paste() to assemble a vector of subscripted α values as character strings suitable for parsing into expressions.

```
set.seed(54321) # make sure we always generate the same data
my.data <-
  data.frame(x = 1:5,
             y = rnorm(5),
             greek.label = paste("alpha[", 1:5, "]", sep = ""))
```

We use as x-axis label, a Greek α character with i as subscript, and in the y-axis label, we have a superscript in the units. For the title we use a character string but for the subtitle a rather complex expression. We create these expressions with function expression().

We label each observation with a subscripted *alpha*. We cannot pass expressions to *geometries* by simply mapping them to the label aesthetic. Instead, we pass character strings that can be parsed into expressions. In other words, character strings, that are written using the syntax of expressions. We need to set parse = TRUE in the call to the *geometry* so that the strings, instead of being plotted as is, are parsed into expressions before the plot is rendered.

```
ggplot(my.data, aes(x, y, label = greek.label)) +
  geom_point() +
  geom_text(angle = 45, hjust = 1.2, parse = TRUE) +
  labs(x = expression(alpha[i]),
       y = expression(Speed~~(m~s^{-1})),
       title = "Using expressions",
       subtitle = expression(sqrt(alpha[1] + frac(beta, gamma))))
```

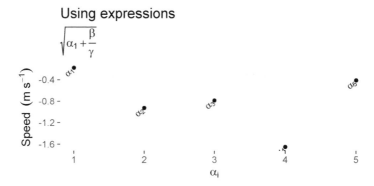

We can also use a character string stored in a variable, and use function `parse()` to parse it in cases where an expression is required as we do here for `subtitle`. In this example we also set tick labels to expressions, taking advantage that `expression()` accepts multiple arguments separated by commas returning a vector of expressions.

```
my_eq.char <- "alpha[i]"
ggplot(my.data, aes(x, y)) +
   geom_point() +
   labs(title = parse(text = my_eq.char)) +
   scale_x_continuous(name = expression(alpha[i]),
                      breaks = c(1,3,5),
                      labels = expression(alpha[1], alpha[3], alpha[5]))
```

A different approach (no example shown) would be to use `parse()` explicitly for each individual label, something that might be needed if the tick labels need to be "assembled" programmatically instead of set as constants.

Differences between `parse()` and `expression()`. Function `parse()` takes as an argument a character string. This is very useful as the character string can be created programmatically. When using `expression()` this is not possible, except for substitution at execution time of the value of variables into the expression. See the help pages for both functions.

Function `expression()` accepts its arguments without any delimiters. Function `parse()` takes a single character string as an argument to be parsed, in which case quotation marks within the string need to be *escaped* (using \"

where a literal " is desired). We can, also in both cases, embed a character string by means of one of the functions `plain()`, `italic()`, `bold()` or `bolditalic()` which also affect the font used. The argument to these functions needs to be a character string delimited by quotation marks if it is not to be parsed.

When using `expression()`, bare quotation marks can be embedded,

```
ggplot(cars, aes(speed, dist)) +
  geom_point() +
  xlab(expression(x[1]*" test"))
```

while in the case of `parse()` they need to be *escaped*,

```
ggplot(cars, aes(speed, dist)) +
  geom_point() +
  xlab(parse(text = "x[1]*\" test\""))
```

and in some cases will be enclosed within a format function.

```
ggplot(cars, aes(speed, dist)) +
  geom_point() +
  xlab(parse(text = "x[1]*italic(\" test\")"))
```

Some additional remarks. If `expression()` is passed multiple arguments, it returns a vector of expressions. Where `ggplot()` expects a single value as an argument, as in the case of axis labels, only the first member of the vector will be used.

```
ggplot(cars, aes(speed, dist)) +
  geom_point() +
  xlab(expression(x[1], " test"))
```

Depending on the location within a expression, spaces maybe ignored, or illegal. To juxtapose elements without adding space use `*`, to explicitly insert white space, use `~`. As shown above, spaces are accepted within quoted text. Consequently, the following alternatives can also be used.

```
xlab(parse(text = "x[1]~~~~\"test\""))
```

```
xlab(parse(text = "x[1]~~~~plain(test)"))
```

However, unquoted white space is discarded.

```
xlab(parse(text = "x[1]*plain( test)"))
```

Finally, it can be surprising that trailing zeros in numeric values appearing within an expression or text to be parsed are dropped. To force the trailing zeros to be retained we need to enclose the number in quotation marks so that it is interpreted as a character string.

```
ggplot(cars, aes(speed, dist)) +
  geom_point() +
  annotate(geom = "text",
           x = rep(6, 3), y = c(90, 100, 110),
           label = c("'1.00'*x^2", "1.00*x^2", "1.01*x^2"), parse = TRUE)
```

Above we used `paste()` to insert values stored in a variable; functions `format()`, `sprintf()`, and `strftime()` allow the conversion into character strings of other values. These functions can be used when creating plots to generate suitable character strings for the `label` *aesthetic* out of numeric, logical, date, time, and even character values. They can be, for example, used to create labels within a call to `aes()`.

```
sprintf("log(%.3f) = %.3f", 5, log(5))
## [1] "log(5.000) = 1.609"
```

```
sprintf("log(%.3g) = %.3g", 5, log(5))
## [1] "log(5) = 1.61"
```

> Study the chunck above. If you are familiar with C or C++ function `sprintf()` will already be familiar to you, otherwise study its help page.
>
> Play with functions `format()`, `sprintf()`, and `strftime()`, formatting different types of data, into character strings of different widths, with different numbers of digits, etc.

It is also possible to substitute the value of variables or, in fact, the result of evaluation, into a new expression, allowing on the fly construction of expressions. Such expressions are frequently used as labels in plots. This is achieved through use of *quoting* and *substitution*.

We use `bquote()` to substitute variables or expressions enclosed in `.( )` by their value. Be aware that the argument to `bquote()` needs to be written as an expression; in this example we need to use a tilde, `~`, to insert a space between words. Furthermore, if the expressions include variables, these will be searched for in the environment rather than in `data`, except within a call to `aes()`.

```
ggplot(cars, aes(speed, dist)) +
  geom_point() +
  labs(title = bquote(Time~zone: .(Sys.timezone())),
       subtitle = bquote(Date: .(as.character(today()))))
       )
```

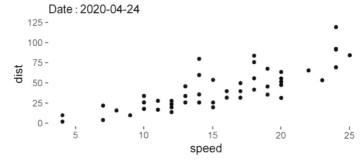

In the case of `substitute()` we supply what is to be used for substitution through a named list.

```r
ggplot(cars, aes(speed, dist)) +
  geom_point() +
  labs(title = substitute(Time~zone: tz, list(tz = Sys.timezone())),
       subtitle = substitute(Date: date, list(date = as.character(today()))))
       )
```

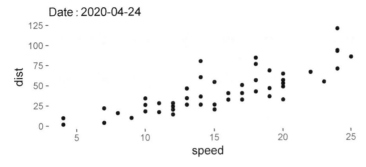

For example, substitution can be used to assemble an expression within a function based on the arguments passed. One case of interest is to retrieve the name of the object passed as an argument, from within a function.

```r
deparse_test <- function(x) {
  print(deparse(substitute(x)))
}

a <- "saved in variable"

deparse_test("constant")
## [1] "\"constant\""

deparse_test(1 + 2)
## [1] "1 + 2"

deparse_test(a)
## [1] "a"
```

> ℹ A new package, 'ggtext', which is not yet in CRAN, provides rich-text (basic HTML and Markdown) support for 'ggplot2', both for annotations and for data visualization. This package provides an alternative to the use of R expressions.

7.13 Creating complex data displays

The grammar of graphics allows one to build and test plots incrementally. In daily use, when creating a completely new plot, it is best to start with a simple design for a plot, `print()` this plot, checking that the output is as expected and the code error-free. Afterwards, one can map additional *aesthetics* and add *geometries* and *statistics* gradually. The final steps are then to add *annotations* and the text or expressions used for titles, and axis and key labels. Another approach is to start with an existing plot and modify it, e.g., by using the same plotting code with different `data` or mapping different variables. When reusing code for a different data set, scale `limits` and `names` are likely to need to be edited.

> Build a graphically complex data plot of your interest, step by step. By step by step, I do not refer to using the grammar in the construction of the plot as earlier, but of taking advantage of this modularity to test intermediate versions in an iterative design process, first by building up the complex plot in stages as a tool in debugging, and later using iteration in the processes of improving the graphic design of the plot and improving its readability and effectiveness.

7.14   Creating sets of plots

Plots to be presented at a given occasion or published as part of the same work need to be consistent in various respects: themes, scales and palettes, annotations, titles and captions. To guarantee this consistency we need to build plots modularly and avoid repetition by assigning names to the "modules" that need to be used multiple times.

7.14.1   Saving plot layers and scales in variables

When creating plots with 'ggplot2', objects are composed using operator + to assemble together the individual components. The functions that create plot layers, scales, etc. are constructors of objects and the objects they return can be stored in variables, and once saved, added to multiple plots at a later time.

We create a plot and save it to variable `myplot` and we separately save the values returned by a call to function `labs()`.

```
myplot <- ggplot(data = mtcars,
                 aes(x = disp, y = mpg,
                     color = factor(cyl))) +
          geom_point()

mylabs <- labs(x = "Engine displacement)",
               y = "Gross horsepower",
               color = "Number of\ncylinders",
               shape = "Number of\ncylinders")
```

We assemble the final plot from the two parts we saved into variables. This is useful when we need to create several plots ensuring that scale `name` arguments are used consistently. In the example above, we saved these names, but the approach can be used for other plot components or lists of components.

> ⚠️ When composing plots with the + operator, the left-hand-side operand must be a `"gg"` object. The left operand is added to the `"gg"` object and the result returned.

```
myplot
myplot + mylabs + theme_bw(16)
myplot + mylabs + theme_bw(16) + ylim(0, NA)
```

We can also save intermediate results.

```
mylogplot <- myplot + scale_y_log10(limits=c(8,55))
mylogplot + mylabs + theme_bw(16)
```

7.14.2   Saving plot layers and scales in lists

If the pieces to be put together do not include a `"gg"` object, we can group them into an R list and save it. When we later add the saved list to a `"gg"` object, the members of the list are added one by one to the plot respecting their order.

```
myparts <- list(mylabs, theme_bw(16))
mylogplot + myparts
```

> 📖 Revise the code you wrote for the "playground" exercise in section 7.13, but this time, pre-building and saving groups of elements that you expect to be useful unchanged when composing a different plot of the same type, or a plot of a different type from the same data.

7.14.3   Using functions as building blocks

When the blocks we assemble need to accept arguments when used, we have to define functions instead of saving plot components to variables. The functions we define, have to return a `"gg"` object, a list of plot components, or a single plot component. The simplest use is to alter some defaults in existing constructor functions returning `"gg"` objects or layers. The ellipsis (...) allows passing named arguments to a nested function. In this case, every single argument passed by name to `bw_ggplot()` will be copied as argument to the nested call to `ggplot()`. Be aware, that supplying arguments by position, is possible only for parameters explicitly included in the definition of the wrapper function,

```
bw_ggplot <- function(...) {
  ggplot(...) +
  theme_bw()
}
```

which could be used as follows.

```
bw_ggplot(data = mtcars,
          aes(x = disp, y = mpg,
          color = factor(cyl))) +
          geom_point()
```

7.15 Generating output files

It is possible, when using RStudio, to directly export the displayed plot to a file using a menu. However, if the file will have to be generated again at a later time, or a series of plots need to be produced with consistent format, it is best to include the commands to export the plot in the script.

In R, files are created by printing to different devices. Printing is directed to a currently open device such a window in RStudio. Some devices produce screen output, others files. Devices depend on drivers. There are both devices that are part of R and additional ones defined in contributed packages.

Creating a file involves opening a device, printing and closing the device in sequence. In most cases the file remains locked until the device is close.

For example when rendering a plot to PDF, Encapsulated Postcript, SVG or other vector graphics formats, arguments passed to `width` and `height` are expressed in inches.

```
fig1 <- ggplot(data.frame(x = -3:3), aes(x = x)) +
  stat_function(fun = dnorm)
pdf(file = "fig1.pdf", width = 8, height = 6)
print(fig1)
dev.off()
```

For Encapsulated Postscript and SVG output, we only need to substitute `pdf()` with `postscript()` or `svg()`, respectively.

```
postscript(file = "fig1.eps", width = 8, height = 6)
print(fig1)
dev.off()
```

In the case of graphics devices for file output in BMP, JPEG, PNG and TIFF bitmap formats, arguments passed to `width` and `height` are expressed in pixels.

```
tiff(file = "fig1.tiff", width = 1000, height = 800)
print(fig1)
dev.off()
```

> ℹ Some graphics devices are part of base-R, and others are implemented in contributed packages. In some cases, there are multiple graphic device available for rendering graphics in a given file format. These devices usually use

different libraries, or have been designed with different aims. These alternative graphic devices can also differ in their function signature, i.e., have differences in the parameters and their names. In cases when rendering fails inexplicably, it can be worthwhile to switch to an alternative graphics device to find out if the problem is in the plot or in the rendering engine.

7.16   Further reading

An in-depth discussion of the many extensions to package 'ggplot2' is outside the scope of this book. Several books describe in detail the use of 'ggplot2', being *ggplot2: Elegant Graphics for Data Analysis* (Wickham and Sievert 2016) the one written by the main author of the package. For inspiration or worked out examples, the book *R Graphics Cookbook* (Chang 2018) is an excellent reference. In depth explanations of the technical aspects of R graphics are available in the book *R Graphics* (Murrell 2019).

8

Data import and export

Most programmers have seen them, and most good programmers realize they've written at least one. They are huge, messy, ugly programs
that should have been short, clean, beautiful programs.

John Bentley
Programming Pearls, 1986

8.1 Aims of this chapter

Base R and the recommended packages (installed by default) include several functions for importing and exporting data. Contributed packages provide both re-
placements for some of these functions and support for several additional file
formats. In the present chapter, I aim at describing both data input and output
covering in detail only the most common "foreign" data formats (those not native
to R).

Data file formats that are foreign to R are not always well defined, making it
necessary to reverse-engineer the algorithms needed to read them. These formats,
even when clearly defined, may be updated by the developers of the foreign software that writes the files. Consequently, developing software to read and write files
using foreign formats can easily result in long, messy, and ugly R scripts. We can
also unwillingly write code that usually works but occasionally fails with specific
files, or even worse, occasionally silently corrupts the imported data. The aim of
this chapter is to provide guidance for finding functions for reading data encoded
using foreign formats, covering both base R, including the 'foreign' package, and
independently contributed packages. Such functions are well tested or validated.

In this chapter you will familiarize yourself with how to exchange data between
R and other applications. The functions save() and load(), and saveRDS() and
readRDS(), all of which save and read data in R's native formats, are described in
sections 2.16.2 and 2.16.3 starting on page 79.

8.2   Introduction

The first step in any data analysis with R is to input or read-in the data. Available sources of data are many and data can be stored or transmitted using various formats, both based on text or binary encodings. It is crucial that data is not altered (corrupted) when read and that in the eventual case of an error, errors are clearly reported. Most dangerous are silent non-catastrophic errors.

The very welcome increase of awareness of the need for open availability of data, makes the output of data from R into well-defined data-exchange formats another crucial step. Consequently, in many cases an important step in data analysis is to export the data for submission to a repository, in addition to publication of the results of the analysis.

Faster internet access to data sources and cheaper random-access memory (RAM) has made it possible to efficiently work with relatively large data sets in R. That R keeps all data in memory (RAM), imposes limits to the size of data R functions can operate on. For data sets large enough not to fit in computer RAM, one can use selective reading of data from flat files, or from databases outside of R.

Some R packages have made it faster to import data saved in the same formats already supported by base R, but in some cases providing weaker guarantees of not corrupting the data than base R. Other contributed packages make it possible to import and export data stored in file formats not supported by base R functions. Some of these formats are subject-area specific while others are in widespread use.

8.3   Packages used in this chapter

```
install.packages(learnrbook::pkgs_ch_data)
```

To run the examples included in this chapter, you need first to load some packages from the library (see section 5.2 on page 163 for details on the use of packages).

```
library(learnrbook)
library(tibble)
library(purrr)
library(wrapr)
library(stringr)
library(dplyr)
library(tidyr)
library(readr)
library(readxl)
library(xlsx)
library(readODS)
library(pdftools)
library(foreign)
library(haven)
```

```
library(xml2)
library(XML)
library(ncdf4)
library(tidync)
library(lubridate)
library(jsonlite)
```

ⓘ Some data sets used in this and other chapters are available in package 'learnrbook'. In addition to the R data objects, we provide files saved in *foreign* formats, which we used in examples on how to import data. The files can be either read from the R library, or from a copy in a local folder. In this chapter we assume the user has copied the folder "extdata" from the package to a working folder.

Copy the files using:

```
pkg.path <- system.file("extdata", package = "learnrbook")
file.copy(pkg.path, ".", overwrite = TRUE, recursive = TRUE)
## [1] TRUE
```

We also make sure the folder used to save data read from the internet, exists.

```
save.path = "./data"
if (!dir.exists(save.path)) {
  dir.create(save.path)
}
```

8.4 File names and operations

We start with the naming of files as it affects data sharing irrespective of the format used for its encoding. The main difficulty is that different operating systems have different rules governing the syntax used for file names and file paths. In many cases, like when depositing data files in a public repository, we need to ensure that file names are valid in multiple operating systems (OSs). If the script used to create the files is itself expected to be OS agnostic, we also need to be careful to query the OS for file names and paths without making assumptions on the naming rules or available OS commands. This is especially important when developing R packages.

⚠ For maximum portability, file names should never contain white-space characters and contain at most one dot. For the widest possible portability, underscores should be avoided using dashes instead. As an example, instead of `my data.2019.csv`, use `my-data-2019.csv`.

R provides functions which help with portability, by hiding the idiosyncrasies of the different OSs from R code. In scripts these functions should be preferred over direct call to OS commands (i.e., using `shell()` or `system()`) whenever possible. As the algorithm needed to extract a file name from a file path is OS specific, R provides functions such as `basename()`, whose implementation is OS specific but from the side of R code behave identically—these functions hide the differences among OSs from the user of R. The chunk below can be expected to work correctly under any OS for which R is available.

```
basename("extdata/my-file.txt")
## [1] "my-file.txt"
```

⚠ While in Unix and Linux folder nesting in file paths is marked with a forward slash character (/), under MS-Windows it is marked with a backslash character (\). Backslash (\) is an escape character in R and interpreted as the start of an embedded special sequence of characters (see section 2.7 on page 39), while in R a forward slash (/) can be used for file paths under any OS, and escaped backslash (\\) is valid only under MS-Windows. Consequently, / should be always preferred to \\ to ensure portability, and is the approach used in this book.

```
basename("extdata/my-file.txt")
## [1] "my-file.txt"

basename("extdata\\my-file.txt")
## [1] "my-file.txt"
```

The complementary function to `basename()` is `dirname()` and extracts the bare path to the containing folder, from a full file path.

```
dirname("extdata/my-file.txt")
## [1] "extdata"
```

Functions `getwd()` and `setwd()` can be used to get the path to the current working directory and to set a directory as current, respectively.

```
# not run
getwd()
```

Function `setwd()` returns the path to the current working directory, allowing us to portably set the working directory to the previous one. Both relative paths (relative to the current working directory), as in the example, or absolute paths (given in full) are accepted as an argument. In mainstream OSs "." indicates the current directory and ".." the directory above the current one.

```
# not run
oldwd <- setwd("..")
getwd()
```

The returned value is always an absolute full path, so it remains valid even if the path to the working directory changes more than once before being restored.

```
# not run
oldwd
setwd(oldwd)
getwd()
```

We can also obtain lists of files and/or directories (= disk folders) portably across OSs.

```
head(list.files())
## [1] "abbrev.sty"       "anscombe.svg"       "appendixes.prj"
## [4] "appendixes.prj.bak" "bits"              "chapters-removed"

head(list.dirs())
## [1] "."               "./.git"           "./.git/hooks"      "./.git/info"
## [5] "./.git/logs"     "./.git/logs/refs"

head(dir())
## [1] "abbrev.sty"       "anscombe.svg"       "appendixes.prj"
## [4] "appendixes.prj.bak" "bits"              "chapters-removed"
```

> The default argument for parameter path is the current working directory, under Windows, Unix, and Linux indicated by ".". Convince yourself that this is indeed the default by calling the functions with an explicit argument. After this, play with the functions trying other existing and non-existent paths in your computer.

> Use parameter full.names with list.files() to obtain either a list of file paths or bare file names. Similarly, investigate how the returned list of files is affected by the argument passed to all.names.

> Compare the behavior of functions dir() and list.dirs(), and try by overriding the default arguments of list.dirs(), to get the call to return the same output as dir() does by default.

Base R provides several functions for portably working with files, and they are listed in the help page for files and in individual help pages. Use help("files") to access the help for this "family" of functions.

```r
if (!file.exists("xxx.txt")) {
  file.create("xxx.txt")
}
## [1] TRUE

file.size("xxx.txt")
## [1] 0

file.info("xxx.txt")
##           size isdir mode                mtime                ctime
## xxx.txt      0 FALSE  666 2020-04-24 02:52:45 2020-04-24 02:52:45
##                          atime exe
## xxx.txt 2020-04-24 02:52:45  no

file.rename("xxx.txt", "zzz.txt")
## [1] TRUE

file.exists("xxx.txt")
## [1] FALSE

file.exists("zzz.txt")
## [1] TRUE

file.remove("zzz.txt")
## [1] TRUE
```

> Function `file.path()` can be used to construct a file path from its components in a way that is portable across OSs. Look at the help page and play with the function to assemble some paths that exist in the computer you are using.

8.5 Opening and closing file connections

Examples in the rest of this chapter use as an argument for the `file` formal parameter literal paths or URLs, and complete the reading or writing operations within the call to a function. Sometimes it is necessary to read or write a text file sequentially, one row or record at a time. In such cases it is most efficient to keep the file open between reads and close the connection only when it is no longer needed. See `help(connections)` for details about the various functions available and their behavior in different OSs. In the next example we open a file connection, read two lines, first the top one with column headers, then in a separate call to `readLines()`, the two lines or records with data, and finally close the connection.

```r
f1 <- file("extdata/not-aligned-ASCII-UK.csv", open = "r") # open for reading
readLines(f1, n = 1L)
## [1] "col1,col2,col3,col4"
```

```
readLines(f1, n = 2L)
## [1] "1.0,24.5,346,ABC" "23.4,45.6,78,Z Y"
```

```
close(f1)
```

When R is used in batch mode, the "files" `stdin`, `stdout` and `stderror` can be opened, and data read from, or written to. These *standard* sources and sinks, so familiar to C programmers, allow the use of R scripts as tools in data pipes coded as shell scripts under Unix and other OSs.

8.6   Plain-text files

In general, text files are the most portable approach to data storage but usually also the least efficient with respect to the size of the file. Text files are composed of encoded characters. This makes them easy to edit with text editors and easy to read from programs written in most programming languages. On the other hand, how the data encoded as characters is arranged can be based on two different approaches: positional or using a specific character as a separator. The positional approach is more concise but almost unreadable to humans as the values run into each other. Reading of data stored using a positional approach requires access to a format definition and was common in FORTRAN and COBOL at the time when punch cards were used to store data. In the case of separators, different separators are in common use. Comma-separated values (CSV) encodings use either a comma or semicolon to separate the fields or columns. Tabulator, or tab-separated values (TSV) use the tab character as a column separator. Sometimes white space is used as a separator, most commonly when all values are to be converted to `numeric`.

 Not all text files are born equal. When reading text files, and *foreign* binary files which may contain embedded text strings, there is potential for their misinterpretation during the import operation. One common source of problems, is that column headers are to be read as R names. As earlier discussed, there are strict rules, such as avoiding spaces or special characters if the names are to be used with the normal syntax. On import, some functions will attempt to sanitize the names, but others not. Most such names are still accessible in R statements, but a special syntax is needed to protect them from triggering syntax errors through their interpretation as something different than variable or function names—in R jargon we say that they need to be quoted.

Some of the things we need to be on the watch for are: 1) Mismatches between the character encoding expected by the function used to read the file, and the encoding used for saving the file—usually because of different locales. 2) Leading or trailing (invisible) spaces present in the character values or column names—which are almost invisible when data frames are printed. 3) Wrongly guessed column classes—a typing mistake affecting a single value in a column, e.g., the wrong kind of decimal marker, prevents the column from being recognized as numeric. 4) Mismatched decimal marker in csv files—the marker depends on the locale (language and sometimes country) settings.

If you encounter problems after import, such as failure of indexing of data frame columns by name, use function names() to get the names printed to the console as a character vector. This is useful because character vectors are always printed with each string delimited by quotation marks making leading and trailing spaces clearly visible. The same applies to use of levels() with factors created with data that might have contained mistakes.

To demonstrate some of these problems, I create a data frame with name sanitation disabled, and in the second statement with sanitation enabled. The first statement is equivalent to the default behavior of functions in package 'readr' and the second is equivalent to the behavior of base R functions. 'readr' prioritizes the integrity of the original data while R prioritizes compatibility with R's naming rules.

```
data.frame(a = 1, "a " = 2, " a" = 3, check.names = FALSE)
##    a a   a
## 1  1 2   3

data.frame(a = 1, "a " = 2, " a" = 3)
##    a a. X.a
## 1  1  2   3
```

An even more subtle case is when characters can be easily confused by the user reading the output: zero and o (a0 vs. ao) or el and one (al vs. a1) can be difficult to distinguish in some fonts. When using encodings capable of storing many character shapes, such as unicode, in some cases two characters with almost identical visual shape may be encoded as different characters.

```
data.frame(al = 1, al = 2, a0 = 3, a0 = 4)
##    al al a0 a0
## 1  1  2  3  4
```

Reading data from a text file can result in very odd-looking values stored in R variables because of a mismatch in encoding, e.g., when a CSV file saved with MS-Excel is silently encoded using 16-bit unicode format, but read as an 8-bit unicode encoded file.

The hardest part of all these problems is to diagnose their origin, as function arguments and working environment options can in most cases be used to force the correct decoding of text files with diverse characteristics, origins and vintages once one knows what is required. One function in the R 'tools' package, which is not exported, can at the time of writing be used to test files for the presence on non-ASCII characters: tools:::showNonASCIIfile(). This function takes as an argument the path to a file.

8.6.1   Base R and 'utils'

Text files containing data in columns can be divided into two broad groups. Those with fixed-width fields and those with delimited fields. Fixed-width fields were especially common in the early days of FORTRAN and COBOL when data storage capacity was very limited. These formats are frequently capable of encoding information using fewer characters than when delimited fields are used. The best way of understanding the differences is with examples. Although in this section we exemplify the use of functions by passing a file name as an argument, URLs, and open file descriptors are also accepted (see section 8.5 on page 298).

In the first example we will read a file with fields solely delimited by ",." This is what is called comma-separated-values (CSV) format which can be read and written with `read.csv()` and `write.csv()`, respectively.

Example file `not-aligned-ASCII-UK.csv` contains:

```
col1,col2,col3,col4
1.0,24.5,346,ABC
23.4,45.6,78,Z Y
```

```
from_csv_a.df <- read.csv("extdata/not-aligned-ASCII-UK.csv")
```

```
sapply(from_csv_a.df, class)
##        col1       col2       col3       col4
##   "numeric"   "numeric"   "integer" "character"

from_csv_a.df[["col4"]]
## [1] "ABC" "Z Y"

levels(from_csv_a.df[["col4"]])
## NULL
```

> 📖 Read the file `not-aligned-ASCII-UK.csv` with function `read.csv2()` instead of `read.csv()`. Although this may look like a waste of time, the point of the exercise is for you to get familiar with R behavior in case of such a mistake. This will help you recognize similar errors when they happen accidentally, which is quite common when files are shared.

Example file `aligned-ASCII-UK.csv` contains comma-separated-values with added white space to align the columns, to make it easier to read by humans. These aligned fields contain leading and trailing white spaces that are included in string values when the file is read.

```
col1, col2, col3, col4
 1.0, 24.5,  346,  ABC
23.4, 45.6,   78,  Z Y
```

Although space characters are read as part of the fields, they are ignored when conversion to numeric takes place. The remaining leading and trailing spaces in character strings are difficult to see when data frames are printed.

```
from_csv_b.df <- read.csv("extdata/aligned-ASCII-UK.csv")
```

Using `levels()` we can more clearly see that the labels of the automatically created factor levels contain leading spaces.

```
sapply(from_csv_b.df, class)
##        col1         col2         col3         col4
##   "numeric"    "numeric"    "integer"  "character"

from_csv_b.df[["col4"]]
## [1] "  ABC" "  Z Y"

levels(from_csv_b.df[["col4"]])
## NULL
```

By default, column names are sanitized but factor levels are not. By consulting the documentation with `help(read.csv)` we discover that by passing an additional argument we can change this default and obtain the data read as desired. Most likely the default has been chosen so that by default data integrity is maintained.

```
from_csv_e.df <- read.csv("extdata/aligned-ASCII-UK.csv", strip.white = TRUE)
sapply(from_csv_e.df, class)
##        col1         col2         col3         col4
##   "numeric"    "numeric"    "integer"  "character"

from_csv_e.df[["col4"]]
## [1] "ABC" "Z Y"

levels(from_csv_e.df[["col4"]])
## NULL
```

The functions from the R 'utils' package by default convert columns containing character strings into factors, as seen above. This default can be changed so that character strings remain as is.

```
from_csv_c.df <- read.csv("extdata/not-aligned-ASCII-UK.csv",
                          stringsAsFactors = FALSE)


sapply(from_csv_c.df, class)
##        col1         col2         col3         col4
##   "numeric"    "numeric"    "integer"  "character"

from_csv_c.df[["col4"]]
## [1] "ABC" "Z Y"
```

Decimal points and exponential notation are allowed for floating point values. In English-speaking locales, the decimal mark is a point, while in many other locales it is a comma. If a comma is used as decimal marker, we can no longer use it as field separator and is usually substituted by a semicolon (;). In such a case we can use `read.csv2()` and `write.csv2`. Furthermore, parameters `dec` and `sep` allow setting them to arbitrary characters. Function `read.table()` does the actual work and functions like `read.csv()` only differ in the default arguments for the different parameters. By default, `read.table()` expects fields to be separated by white space (one or more spaces, tabs, new lines, or carriage return). Strings with embedded spaces need to be quoted in the file as shown below.

```
col1 col2 col3 col4
 1.0 24.5  346 ABC
23.4 45.6   78 "Z Y"
```

```
from_txt_b.df <- read.table("extdata/aligned-ASCII.txt", header = TRUE)
```

```
sapply(from_txt_b.df, class)
##        col1        col2        col3        col4
##   "numeric"   "numeric"   "integer" "character"
```

```
from_txt_b.df[["col4"]]
## [1] "ABC" "Z Y"
```

```
levels(from_txt_b.df[["col4"]])
## NULL
```

With a fixed-width format, no delimiters are needed. Decoding is based solely on the position of the characters in the line or record. A file like this cannot be interpreted without a description of the format used for saving the data. Files containing data stored in *fixed width format* can be read with function read.fwf(). Records for a single observation can be stored in a single or multiple lines. In either case, each line has fields of different but fixed known widths.

Function read.fortran() is a wrapper on read.fwf() that accepts format definitions similar to those used in FORTRAN. One particularity of FORTRAN *formatted data transfer* is that the decimal marker can be omitted in the saved file and its position specified as part of the format definition, a trick used to make text files (or stacks of punch cards!) smaller. Modern versions of FORTRAN support reading from and writing to other formats like those using field delimiters described above.

```
 10245346ABC
234456 78Z Y
```

```
from_fwf_a.df <- read.fortran("extdata/aligned-ASCII.fwf",
                              format = c("2F3.1", "F3.0", "A3"),
                              col.names = c("col1", "col2", "col3", "col4"))
```

```
sapply(from_fwf_a.df, class)
##        col1        col2        col3        col4
##   "numeric"   "numeric"   "numeric" "character"
```

```
from_fwf_a.df[["col4"]]
## [1] "ABC" "Z Y"
```

☕ The file reading functions described above share with `read.table()` the same parameters. In addition to those described above, other frequently useful parameters are `skip` and `n`, which can be used to skip lines at the top of a file and limit the number of lines (or records) to read; `header`, which accepts a logical argument indicating if the fields in the first text line read should be decoded as column names rather than data; `na.strings`, to which can be passed a character vector with strings to be interpreted as NA; and `colClasses`, which provides control of the conversion of the fields to R classes and possibly skipping some columns altogether. All these parameters are described in the corresponding help pages.

📊 In reality `read.csv()`, `read.csv2()` and `read.table()` are the same function with different default arguments to several of their parameters. Study the help page, and by passing suitable arguments, make `read.csv()` behave like `read.table()`, then make `read.table()` behave like `read.csv2()`.

☕ We can read a text file as character strings, without attempting to decode them. This is occasionally useful, such as when we do the decoding as part of our own script. In this case, the function to use is `readLines()`. The returned value is a character vector in which each member string corresponds to one line or record in the file, with the end-of-line markers stripped (see example in section 8.5 on page 298).

Next we give one example of the use of a *write* function matching one of the *read* functions described above. The `write.csv()` function takes as an argument a data frame, or an object that can be coerced into a data frame, converts it to character strings, and saves them to a text file. We first create the data frame that we will write to disk.

```r
my.df <- data.frame(x = 1:5, y = 5:1 / 10, z = letters[1:5])
```

We write `my.df` to a CSV file suitable for an English language locale, and then display its contents.

```r
write.csv(my.df, file = "my-file1.csv", row.names = FALSE)
file.show("my-file1.csv", pager = "console")
```

```
"x","y","z"
1,0.5,"a"
2,0.4,"b"
3,0.3,"c"
4,0.2,"d"
5,0.1,"e"
```

In most cases setting, as above, `row.names` = FALSE when writing a CSV file will help when it is read. Of course, if row names do contain important information, such as gene tags, you cannot skip writing the row names to the file unless you first copy these data into a column in the data frame. (Row names are stored separately as an attribute in `data.frame` objects, see section 2.15 on page 77 for details.)

Write the data frame `my.df` into text files with functions `write.csv2()` and `write.table()` instead of `read.csv()` and display the files.

Function `cat()` takes R objects and writes them after conversion to character strings to the console or a file, inserting one or more characters as separators, by default, a space. This separator can be set through parameter `sep`. In our example we set `sep` to a new line (entered as the escape sequence "\n").

```
my.lines <- c("abcd", "hello world", "123.45")
cat(my.lines, file = "my-file2.txt", sep = "\n")
file.show("my-file2.txt", pager = "console")
```

```
abcd
hello world
123.45
```

8.6.2 'readr'

Package 'readr' is part of the 'tidyverse' suite. It defines functions that allow faster input and output, and have different default behavior. Contrary to base R functions, they are optimized for speed, but may sometimes wrongly decode their input and rarely even silently do this. Base R functions do less *guessing*, e.g., the delimiters must be supplied as arguments. The 'readr' functions guess more properties of the text file format; in most cases they succeed, which is very handy, but occasionally they fail. Automatic guessing can be overridden by passing arguments and this is recommended for scripts that may be reused to read different files in the future. Another important advantage is that these functions read character strings formatted as dates or times directly into columns of class POSIXct. All `write` functions defined in 'readr' have an `append` parameter, which can be used to change the default behavior of overwriting an existing file with the same name, to appending the output at its end.

Although in this section we exemplify the use of these functions by passing a file name as an argument, as is the case with R native functions, URLs, and open file descriptors are also accepted (see section 8.5 on page 298). Furthermore, if the file name ends in a tag recognizable as indicating a compressed file format, the file will be uncompressed on the fly.

> ⚠ The names of functions "equivalent" to those described in the previ-
> ous section have names formed by replacing the dot with an underscore, e.g.,
> read_csv() ≈ read.csv(). The similarity refers to the format of the files read,
> but not the order, names, or roles of their formal parameters. For example,
> function read_table() has a slightly different behavior than read.table(), al-
> though they both read fields separated by white space. Other aspects of the
> default behavior are also different, for example 'readr' functions do not con-
> vert columns of character strings into factors and row names are not set in the
> returned tibble, which inherits from data.frame, but is not fully compatible
> (see section 6.4.2 on page 182).

As we can see in this first example, these functions also report to the console
the specifications of the columns, which is important when these are guessed from
the file contents, or part of it.

```
read_csv(file = "extdata/aligned-ASCII-UK.csv")

## Parsed with column specification:
## cols(
##   col1 = col_double(),
##   col2 = col_double(),
##   col3 = col_double(),
##   col4 = col_character()
## )
## # A tibble: 2 x 4
##     col1  col2  col3 col4
##    <dbl> <dbl> <dbl> <chr>
## 1   1    24.5   346 ABC
## 2  23.4  45.6    78 Z Y
```

```
read_csv(file = "extdata/not-aligned-ASCII-UK.csv")

## Parsed with column specification:
## cols(
##   col1 = col_double(),
##   col2 = col_double(),
##   col3 = col_double(),
##   col4 = col_character()
## )
## # A tibble: 2 x 4
##     col1  col2  col3 col4
##    <dbl> <dbl> <dbl> <chr>
## 1   1    24.5   346 ABC
## 2  23.4  45.6    78 Z Y
```

Function read_table(), differently to read.table(), retains quotes as part of
read character strings.

```
read_table(file = "extdata/aligned-ASCII.txt")

## Parsed with column specification:
## cols(
```

```
##   col1 = col_double(),
##   col2 = col_double(),
##   col3 = col_double(),
##   col4 = col_character()
## )
```

Because of the misaligned fields in file "not-aligned-ASCII.txt", we need to use read_table2(), which allows misalignment of fields, like read.table(), instead of read_table(), which expects vertically aligned fields across rows. However, in this case the embedded space character in the quoted string is misinterpreted and part of the string dropped with a warning.

```
read_table2(file = "extdata/not-aligned-ASCII.txt")
```

```
## Parsed with column specification:
## cols(
##   col1 = col_double(),
##   col2 = col_double(),
##   col3 = col_double(),
##   col4 = col_character()
## )
## Warning: 1 parsing failure.
## row col   expected    actual                                file
##   2  --  4 columns 5 columns 'extdata/not-aligned-ASCII.txt'
## # A tibble: 2 x 4
##      col1  col2  col3 col4
##     <dbl> <dbl> <dbl> <chr>
## 1     1   24.5   346 "ABC"
## 2  23.4  45.6    78 "\"Z"
```

Function read_delim() with space as the delimiter needs to be used.

```
read_delim(file = "extdata/not-aligned-ASCII.txt", delim = " ")
```

```
## Parsed with column specification:
## cols(
##   col1 = col_double(),
##   col2 = col_double(),
##   col3 = col_double(),
##   col4 = col_character()
## )
## # A tibble: 2 x 4
##      col1  col2  col3 col4
##     <dbl> <dbl> <dbl> <chr>
## 1     1   24.5   346 ABC
## 2  23.4  45.6    78 Z Y
```

Function read_tsv() reads files encoded with the tab character as the delimiter, and read_fwf() reads files with fixed width fields. There is, however, no equivalent to read.fortran(), supporting implicit decimal points.

> 📖 Use the "wrong" `read_` functions to read the example files used above and/or your own files. As mentioned earlier, forcing errors will help you learn how to diagnose when such errors are caused by coding mistakes. In this case, as wrongly read data are not always accompanied by error or warning messages, carefully check the returned tibbles for misread data values.

> ☕ The functions from R's 'utils' read the whole file as text before attempting to guess the class of the columns or their alignment. This is reliable but slow for very large text files. The functions from 'readr' read only the top 1000 lines by default for guessing, and then rather blindly read the whole files assuming that the guessed properties also apply to the remainder of the file. This is more efficient, but somehow risky. In earlier versions of 'readr', a typical failure to correctly decode fields was when numbers are in increasing order and the field widths continue increasing in the lines below those used for guessing, but this case seems to be, at the time of writing correctly, handled. It also means that in cases when an individual value after `guess_max` lines cannot be converted to numeric, instead of returning a column of character strings as base R functions, this value is encoded as a numeric NA with a warning. To demonstrate this we will drastically reduce `guess_max` from its default so that we can use an example file only a few lines in length.
>
> ```
> read_table2(file = "extdata/miss-aligned-ASCII.txt")
> ```
>
> ```
> ## Parsed with column specification:
> ## cols(
> ##   col1 = col_character(),
> ##   col2 = col_double(),
> ##   col3 = col_double(),
> ##   col4 = col_character()
> ## )
> ## # A tibble: 4 x 4
> ##    col1   col2  col3 col4
> ##    <chr> <dbl> <dbl> <chr>
> ## 1 1.0    24.5   346 ABC
> ## 2 2.4    45.6    78 XYZ
> ## 3 20.4   45.6    78 XYZ
> ## 4 a      20    2500 abc
> ```

```
read_table2(file = "extdata/miss-aligned-ASCII.txt", guess_max = 3L)

## Parsed with column specification:
## cols(
##   col1 = col_double(),
##   col2 = col_double(),
##   col3 = col_double(),
##   col4 = col_character()
## )
## Warning: 1 parsing failure.
## row  col expected actual                              file
##   4 col1 a double      a 'extdata/miss-aligned-ASCII.txt'
## # A tibble: 4 x 4
##      col1  col2  col3 col4
##     <dbl> <dbl> <dbl> <chr>
## 1     1    24.5   346 ABC
## 2    2.4   45.6    78 XYZ
## 3   20.4   45.6    78 XYZ
## 4    NA    20    2500 abc
```

The `write_` functions from 'readr' are the counterpart to `write.` functions from 'utils'. In addition to the expected `write_csv()`, `write_csv2()`, `write_tsv()` and `write_delim()`, 'readr' provides functions that write MS-Excel-friendly CSV files. We demonstrate here the use of `write_excel_csv()` to produce a text file with comma-separated fields suitable for import into MS-Excel.

```
write_excel_csv(my.df, path = "my-file6.csv")
file.show("my-file6.csv", pager = "console")
```

That saves a file containing the following text:

```
x,y,z
1,0.5,a
2,0.4,b
3,0.3,c
4,0.2,d
5,0.1,e
```

> Compare the output from `write_excel_csv()` and `write_csv()`. What is the difference? Does it matter when you import the written CSV file into Excel (the version you are using, with the locale settings of your computer)?

The pair of functions `read_lines()` and `write_lines()` read and write character vectors without conversion, similarly to base R `readLines()` and `writeLines()`. Functions `read_file()` and `write_file()` read and write the contents of a whole text file into, and from, a single character string. Functions `read_file()` and `write_file()` can also be used with raw vectors to read and write binary files or text files of unknown encoding.

The contents of the whole file are returned as a character vector of length one, with the embedded new line markers. We use cat() to print it so these new line characters force the start of a new print-out line.

```
one.str <- read_file(file = "extdata/miss-aligned-ASCII.txt")
length(one.str)
## [1] 1

cat(one.str)
## col1   col2 col3 col4

## 1.0    24.5  346 ABC

## 2.4    45.6   78 XYZ

## 20.4    45.6   78 XYZ

##   a     20      2500 abc
```

 Use write_file() to write a file that can be read with read_csv().

8.7 XML and HTML files

XML files contain text with special markup. Several modern data exchange formats are based on the XML standard (see https://www.w3.org/TR/xml/) which uses schemas for flexibility. Schemas define specific formats, allowing reading of formats not specifically targeted during development of the read functions. Even the modern XHTML standard used for web pages is based on such schemas, while HTML only differs slightly in its syntax.

8.7.1 'xml2'

Package 'xml2' provides functions for reading and parsing XTML and HTML files. This is a vast subject, of which I will only give a brief example.

We first read a web page with function read_html(), and explore its structure.

```
web_page <- read_html("http://r4photobiology.info/R/index.html")
html_structure(web_page)
## <html>
##   <head>
##     <title>
##       {text}
##     <meta [name, content]>
##     <meta [name, content]>
##     <meta [name, content]>
##   <body>
```

```
##        {text}
##        <hr>
##        <h1>
##          {text}
##        {text}
##        <hr>
##        <p>
##          {text}
##          <a [href]>
##            {text}
##          {text}
##        {text}
##        <p>
##          {text}
##          <a [href]>
##            {text}
##          {text}
##        {text}
##        <address>
##          {text}
##        {text}
```

Next we extract the text from its `title` attribute, using functions `xml_find_all()` and `xml_text()`.

```
xml_text(xml_find_all(web_page, ".//title"))
## [1] "Suite of R packages for photobiology"
```

The functions defined in this package can be used to "harvest" data from web pages, but also to read data from files using formats that are defined through XML schemas.

8.8   GPX files

GPX (GPS Exchange Format) files use an XML scheme designed for saving and exchanging data from geographic positioning systems (GPS). There is some variation on the variables saved depending on the settings of the GPS receiver. The example data used here is from a Transmeta BT747 GPS logger. The example below reads the data into a `tibble` as character strings. For plotting, the character values representing numbers and dates would need to be converted to numeric and datetime (`POSIXct`) values, respectively. In the case of plotting tracks on a map, it is preferable to use package 'sf' to import the tracks directly from the .gpx file into a layer (use of the dot pipe operator is described in section 6.5 on page 187).

```
xmlTreeParse(file = "extdata/GPSDATA.gpx", useInternalNodes = TRUE) %.>%
  xmlRoot(x = .) %.>%
  xmlToList(node = .)[["trk"]] %.>%
  unlist(x = .[names(.) == "trkseg"], recursive = FALSE) %.>%
  map_df(.x = ., .f = function(x) as_tibble(x = t(x = unlist(x = x))))
## # A tibble: 199 x 7
##   time          speed name      type  fix   .attrs.lat .attrs.lon
```

```
## * <chr>              <chr> <chr>                    <chr> <chr> <chr>        <chr>
## 1 2018-12-08T23:09~ 0.03~ trkpt-2018-12-08T23~ T     3d    -34.912071 138.660595
## 2 2018-12-08T23:09~ 0.08~ trkpt-2018-12-08T23~ T     3d    -34.912067 138.660543
## 3 2018-12-08T23:09~ 0.01~ trkpt-2018-12-08T23~ T     3d    -34.912102 138.660554
## # ... with 196 more rows
```

I have passed all arguments by name to make explicit how this pipe works. See section 6.5 on page 187 for details on the use of the pipe and dot-pipe operators.

> To understand what data transformation takes place in each statement of this pipe, start by executing the first statement by itself, excluding the dot-pipe operator, and continue adding one statement at a time, and at each step check the returned value and look out for what has changed from the previous step.

8.9 Worksheets

Microsoft Office, Open Office and Libre Office are the most frequently used suites containing programs based on the worksheet paradigm. There is available a standardized file format for exchange of worksheet data, but it does not support all the features present in native file formats. We will start by considering MS-Excel. The file format used by MS-Excel has changed significantly over the years, and old formats tend to be less well supported by available R packages and may require the file to be updated to a more modern format with MS-Excel itself before import into R. The current format is based on XML and relatively simple to decode, whereas older binary formats are more difficult. Worksheets contain code as equations in addition to the actual data. In all cases, only values entered as such or those computed by means of the embedded equations can be imported into R rather than the equations themselves.

8.9.1 CSV files as middlemen

If we have access to the original software used for creating a worksheet or workbook, then exporting worksheets to text files in CSV format and importing them into R using the functions described in sections 8.6 and 8.6.2 starting on pages 299 and 305 provides a broadly compatible route for importing data—with the caveat that we should take care that delimiters and decimal marks match the expectations of the functions used. This approach is not ideal from the perspective of having to recreate intermediate files. A better approach is, when feasible, to import the data directly from the workbook or worksheets into R.

8.9.2 'readxl'

Package 'readxl' supports reading of MS-Excel workbooks, and selecting worksheets and regions within worksheets specified in ways similar to those used by

MS-Excel itself. The interface is simple, and the package easy to install. We will import a file that in MS-Excel looks like the screen capture below.

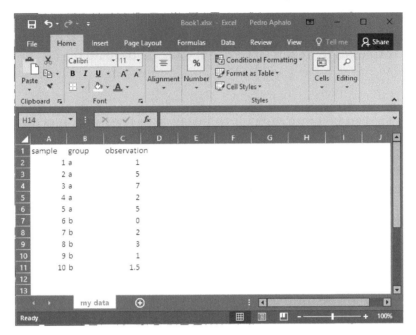

We first list the sheets contained in the workbook file with excel_sheets().

```
sheets <- excel_sheets("extdata/Book1.xlsx")
sheets
## [1] "my data"
```

In this case, the argument passed to sheet is redundant, as there is only a single worksheet in the file. It is possible to use either the name of the sheet or a positional index (in this case 1 would be equivalent to "my data"). We use function read_excel() to import the worksheet. Being part of the 'tidyverse' the returned value is a tibble and character columns are returned as is.

```
Book1.df <- read_excel("extdata/Book1.xlsx", sheet = "my data")
Book1.df
## # A tibble: 10 x 3
##    sample group observation
##     <dbl> <chr>       <dbl>
## 1       1 a               1
## 2       2 a               5
## 3       3 a               7
## # ... with 7 more rows
```

We can also read a region instead of the whole worksheet.

```
Book1_region.df <- read_excel("extdata/Book1.xlsx", sheet = "my data", range = "A1:B8")
Book1_region.df
## # A tibble: 7 x 2
##    sample group
##     <dbl> <chr>
## 1       1 a
```

```
## 2        2 a
## 3        3 a
## # ... with 4 more rows
```

Of the remaining arguments, the most useful ones have the same names and play similar roles as in 'readr' (see section 8.6.2 on page 305).

8.9.3 'xlsx'

Package 'xlsx' can be more difficult to install as it uses Java functions to do the actual work. However, it is more comprehensive, with functions both for reading and writing MS-Excel worksheets and workbooks, in different formats including the older binary ones. Similar to 'readr' it allows selected regions of a worksheet to be imported.

Here we use function read.xlsx(), indexing the worksheet by name. The returned value is a data frame, and following the expectations of R package 'utils', character columns are converted into factors by default.

```
Book1_xlsx.df <- read.xlsx("extdata/Book1.xlsx", sheetName = "my data")
Book1_xlsx.df
##     sample group observation
## 1        1     a         1.0
## 2        2     a         5.0
## 3        3     a         7.0
## 4        4     a         2.0
## 5        5     a         5.0
## 6        6     b         0.0
## 7        7     b         2.0
## 8        8     b         3.0
## 9        9     b         1.0
## 10      10     b         1.5
```

With function write.xlsx() we can write data frames out to Excel worksheets and even append new worksheets to an existing workbook.

```
set.seed(456321)
my.data <- data.frame(x = 1:10, y = letters[1:10])
write.xlsx(my.data, file = "extdata/my-data.xlsx", sheetName = "first copy")
write.xlsx(my.data, file = "extdata/my-data.xlsx", sheetName = "second copy", append = TRUE)
```

When opened in Excel, we get a workbook containing two worksheets, named using the arguments we passed through sheetName in the code chunk above.

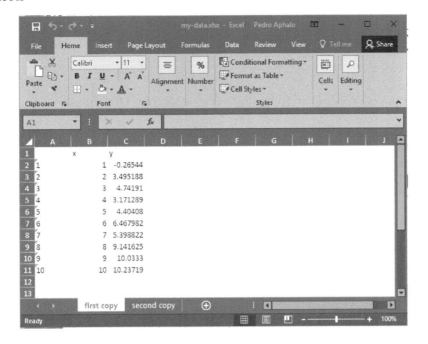

If you have some worksheet files available, import them into R to get a feel for how the way in which data is organized in the worksheets affects how easy or difficult it is to import them into R.

8.9.4 'readODS'

Package 'readODS' provides functions for reading data saved in files that follow the *Open Documents Standard*. Function read_ods() has a similar but simpler user interface to that of read_excel() and reads one worksheet at a time, with support only for skipping top rows. The value returned is a data frame.

```
ods.df <- read_ods("extdata/Book1.ods", sheet = 1)

## Parsed with column specification:
## cols(
##   sample = col_double(),
##   group = col_character(),
##   observation = col_double()
## )


ods.df
##   sample group observation
## 1      1     a         1.0
## 2      2     a         5.0
## 3      3     a         7.0
## 4      4     a         2.0
## 5      5     a         5.0
## 6      6     b         0.0
```

```
## 7        7      b        2.0
## 8        8      b        3.0
## 9        9      b        1.0
## 10      10      b        1.5
```

Function `write_ods()` writes a data frame into an ODS file.

8.10   Statistical software

There are two different comprehensive packages for importing data saved from other statistical programs such as SAS, Statistica, SPSS, etc. The longtime "standard" is package 'foreign' included in base R, and package 'haven' is a newer contributed extension. In the case of files saved with old versions of statistical programs, functions from 'foreign' tend to be more robust than those from 'haven'.

8.10.1   'foreign'

Functions in package 'foreign' allow us to import data from files saved by several statistical analysis programs, including SAS, Stata, SPPS, Systat, Octave among others, and a function for writing data into files with formats native to SAS, Stata, and SPPS. R documents the use of these functions in detail in the *R Data Import/Export* manual. As a simple example, we use function `read.spss()` to read a `.sav` file, saved a few years ago with the then current version of SPSS. We display only the first six rows and seven columns of the data frame, including a column with dates, which appears as numeric.

```
my_spss.df <- read.spss(file = "extdata/my-data.sav", to.data.frame = TRUE)

## re-encoding from UTF-8

my_spss.df[1:6, c(1:6, 17)]
##    block        treat mycotreat water1 pot harvest harvest_date
## 1      0 Watered, EM          1      1  14       1  13653705600
## 2      0 Watered, EM          1      1  52       1  13653705600
## 3      0 Watered, EM          1      1 111       1  13653705600
## 4      0 Watered, EM          1      1 127       1  13653705600
## 5      0 Watered, EM          1      1 230       1  13653705600
## 6      0 Watered, EM          1      1 258       1  13653705600
```

A second example, this time with a simple `.sav` file saved 15 years ago.

```
thiamin.df <- read.spss(file = "extdata/thiamin.sav", to.data.frame = TRUE)
head(thiamin.df)
##    THIAMIN CEREAL
## 1      5.2  wheat
## 2      4.5  wheat
## 3      6.0  wheat
## 4      6.1  wheat
## 5      6.7  wheat
## 6      5.8  wheat
```

Another example, for a Systat file saved on an PC more than 20 years ago, and read with `read.systat()`.

```
my_systat.df <- read.systat(file = "extdata/BIRCH1.SYS")
head(my_systat.df)
##   CONT DENS BLOCK SEEDL VITAL BASE ANGLE HEIGHT DIAM
## 1    1    1     1     2    44    2     0      1   53
## 2    1    1     1     2    41    2     1      2   70
## 3    1    1     1     2    21    2     0      1   65
## 4    1    1     1     2    15    3     0      1   79
## 5    1    1     1     2    37    3     0      1   71
## 6    1    1     1     2    29    2     1      1   43
```

Not all functions in 'foreign' return data frames by default, but all of them can be coerced to do so.

8.10.2 'haven'

Package 'haven' is less ambitious with respect to the number of formats supported, or their vintages, providing read and write functions for only three file formats: SAS, Stata and SPSS. On the other hand, 'haven' provides flexible ways to convert the different labeled values that cannot be directly mapped to R modes. They also decode dates and times according to the idiosyncrasies of each of these file formats. In cases when the imported file contains labeled values the returned `tibble` object needs some additional attention from the user. Labeled numeric columns in SPSS are not necessarily equivalent to factors, although they sometimes are. Consequently, conversion to factors cannot be automated and must be done manually in a separate step.

We can use function `read_sav()` to import a `.sav` file saved by a recent version of SPSS. As in the previous section, we display only the first six rows and seven columns of the data frame, including a column `treat` containing a labeled numeric vector and `harvest_date` with dates encoded as R date values.

```
my_spss.tb <- read_sav(file = "extdata/my-data.sav")
my_spss.tb[1:6, c(1:6, 17)]
## # A tibble: 6 x 7
##   block            treat mycotreat water1   pot harvest harvest_date
##   <dbl>        <dbl+lbl>     <dbl> <dbl> <dbl>   <dbl> <date>
## 1     0 1 [Watered, EM]         1     1    14       1 2015-06-15
## 2     0 1 [Watered, EM]         1     1    52       1 2015-06-15
## 3     0 1 [Watered, EM]         1     1   111       1 2015-06-15
## # ... with 3 more rows
```

In this case, the dates are correctly decoded.

Next, we import an SPSS's `.sav` file saved 15 years ago.

```
thiamin.tb <- read_sav(file = "extdata/thiamin.sav")
thiamin.tb
## # A tibble: 24 x 2
##   THIAMIN    CEREAL
##     <dbl> <dbl+lbl>
## 1     5.2 1 [wheat]
## 2     4.5 1 [wheat]
## 3     6   1 [wheat]
## # ... with 21 more rows
```

```
thiamin.tb <- as_factor(thiamin.tb)
thiamin.tb
## # A tibble: 24 x 2
##    THIAMIN CEREAL
##      <dbl> <fct>
## 1      5.2 wheat
## 2      4.5 wheat
## 3      6   wheat
## # ... with 21 more rows
```

> 📊 Compare the values returned by different read functions when applied to the same file on disk. Use names(), str() and class() as tools in your exploration. If you are brave, also use attributes(), mode(), dim(), dimnames(), nrow() and ncol().

> 📊 If you use or have in the past used other statistical software or a general-purpose language like Python, look for some old files and import them into R.

8.11 NetCDF files

In some fields, including geophysics and meteorology, NetCDF is a very common format for the exchange of data. It is also used in other contexts in which data is referenced to a grid of locations, like with data read from Affymetrix microarrays used to study gene expression. NetCDF files are binary but use a format that allows the storage of metadata describing each variable together with the data itself in a well-organized and standardized format, which is ideal for exchange of moderately large data sets measured on a spatial or spatio-temporal grid.

Officially described as follows:

> NetCDF is a set of software libraries [from Unidata] and self-describing, machine-independent data formats that support the creation, access, and sharing of array-oriented scientific data.

As sometimes NetCDF files are large, it is good that it is possible to selectively read the data from individual variables with functions in packages 'ncdf4' or 'RNetCDF'. On the other hand, this implies that contrary to other data file reading operations, reading a NetCDF file is done in two or more steps—i.e., opening the file, reading metadata describing the variables and spatial grid, and finally reading the data of interest.

8.11.1 'ncdf4'

Package 'ncdf4' supports reading of files using netCDF version 4 or earlier formats. Functions in 'ncdf4' not only allow reading and writing of these files, but also their modification.

We first read metadata to obtain an index of the file contents, and in additional steps, read a subset of the data. With print() we can find out the names and characteristics of the variables and attributes. In this example, we read long-term averages for potential evapotranspiration (PET).

We first open a connection to the file with function nc_open().

```
meteo_data.nc <- nc_open("extdata/pevpr.sfc.mon.ltm.nc")
str(meteo_data.nc, max.level = 1)
## List of 14
##  $ filename   : chr "extdata/pevpr.sfc.mon.ltm.nc"
##  $ writable   : logi FALSE
##  $ id         : int 65536
##  $ safemode   : logi FALSE
##  $ format     : chr "NC_FORMAT_NETCDF4_CLASSIC"
##  $ is_GMT     : logi FALSE
##  $ groups     :List of 1
##  $ fqgn2Rindex:List of 1
##  $ ndims      : num 4
##  $ natts      : num 8
##  $ dim        :List of 4
##  $ unlimdimid : num -1
##  $ nvars      : num 3
##  $ var        :List of 3
##  - attr(*, "class")= chr "ncdf4"
```

> Increase max.level in the call to str() above and study the connection object stores information on the dimensions and for each data variable. You can also print(meteo_data.nc) for a more complete printout once you have understood the structure of the object.

The dimensions of the array data are described with metadata, in our examples mapping indexes to a grid of latitudes and longitudes and into a time vector as a third dimension. The dates are returned as character strings. We get here the variables one at a time with function ncvar_get().

```
time.vec <- ncvar_get(meteo_data.nc, "time")
head(time.vec)
## [1] -657073 -657042 -657014 -656983 -656953 -656922

longitude <-  ncvar_get(meteo_data.nc, "lon")
head(longitude)
## [1] 0.000 1.875 3.750 5.625 7.500 9.375

latitude <- ncvar_get(meteo_data.nc, "lat")
head(latitude)
## [1] 88.5420 86.6531 84.7532 82.8508 80.9473 79.0435
```

The `time` vector is rather odd, as it contains only monthly data as these are long-term averages, but expressed as days from 1800-01-01 corresponding to the first day of each month of year 1. We use package 'lubridate' for the conversion.

We construct a `tibble` object with PET values for one grid point, taking advantage of the *recycling* of short vectors.

```
pet.tb <-
    tibble(time = ncvar_get(meteo_data.nc, "time"),
           month = month(ymd("1800-01-01") + days(time)),
           lon = longitude[6],
           lat = latitude[2],
           pet = ncvar_get(meteo_data.nc, "pevpr")[6, 2, ]
           )
pet.tb
## # A tibble: 12 x 5
##        time month   lon   lat   pet
##       <dbl> <dbl> <dbl> <dbl> <dbl>
## 1 -657073      12  9.38  86.7  4.28
## 2 -657042       1  9.38  86.7  5.72
## 3 -657014       2  9.38  86.7  4.38
## # ... with 9 more rows
```

If we want to read in several grid points, we can use several different approaches. However, the order of nesting of dimensions can make adding the dimensions as columns error prone. It is much simpler to use package 'tidync' described next.

8.11.2 'tidync'

Package 'tidync' provides functions that make it easier to extract subsets of the data from an NetCDF file. We start by doing the same operations as in the examples for 'ncdf4'.

We open the file creating an object and simultaneously activating the first grid.

```
meteo_data.tnc <- tidync("extdata/pevpr.sfc.mon.ltm.nc")
meteo_data.tnc
##
## Data Source (1): pevpr.sfc.mon.ltm.nc ...
##
## Grids (5) <dimension family> : <associated variables>
##
## [1]   D0,D1,D2 : pevpr, valid_yr_count   **ACTIVE GRID** ( 216576 values per variable)
## [2]   D3,D2    : climatology_bounds
## [3]   D0       : lon
## [4]   D1       : lat
## [5]   D2       : time
##
## Dimensions 4 (3 active):
##
##   dim   name  length    min     max start count    dmin     dmax unlim coord_dim
##   <chr> <chr>  <dbl>   <dbl>   <dbl> <int> <int>    <dbl>    <dbl> <lgl> <lgl>
## 1 D0    lon      192 0.      3.58e2     1   192 0.       3.58e2 FALSE TRUE
## 2 D1    lat       94 -8.85e1 8.85e1     1    94 -8.85e1 8.85e1 FALSE TRUE
## 3 D2    time      12 -6.57e5 -6.57e5    1    12 -6.57e5 -6.57e5 FALSE TRUE
##
## Inactive dimensions:
```

```
##
##    dim    name  length    min    max unlim coord_dim
##   <chr>  <chr>   <dbl>  <dbl>  <dbl> <lgl>    <lgl>
## 1 D3     nbnds       2      1      2 FALSE    FALSE
```

```
hyper_dims(meteo_data.tnc)
## # A tibble: 3 x 7
##   name   length start count    id unlim coord_dim
## * <chr>   <dbl> <int> <int> <int> <lgl>     <lgl>
## 1 lon       192     1   192     0 FALSE      TRUE
## 2 lat        94     1    94     1 FALSE      TRUE
## 3 time       12     1    12     2 FALSE      TRUE
```

```
hyper_vars(meteo_data.tnc)
## # A tibble: 2 x 6
##      id name               type     ndims natts dim_coord
##   <int> <chr>              <chr>    <int> <int>     <lgl>
## 1     4 pevpr              NC_FLOAT     3    14     FALSE
## 2     5 valid_yr_count     NC_FLOAT     3     4     FALSE
```

We extract a subset of the data into a tibble in long (or tidy) format, and add the months using a pipe operator from 'wrapr' and methods from 'dplyr'.

```
hyper_tibble(meteo_data.tnc,
             lon = signif(lon, 1) == 9,
             lat = signif(lat, 2) == 87) %.>%
  mutate(.data = ., month = month(ymd("1800-01-01") + days(time))) %.>%
  select(.data = ., -time)
## # A tibble: 12 x 5
##   pevpr valid_yr_count    lon    lat month
##   <dbl>          <dbl>  <dbl>  <dbl> <dbl>
## 1  4.28        1.19e-39   9.38   86.7    12
## 2  5.72        1.19e-39   9.38   86.7     1
## 3  4.38        1.29e-39   9.38   86.7     2
## # ... with 9 more rows
```

In this second example, we extract data for all grid points along latitudes. To achieve this we need only to omit the test for `lat` from the chuck above. The tibble is assembled automatically and columns for the active dimensions added. The decoding of the months remains unchanged.

```
hyper_tibble(meteo_data.tnc,
             lon = signif(lon, 1) == 9) %.>%
  mutate(.data = ., month = month(ymd("1800-01-01") + days(time))) %.>%
  select(.data = ., -time)
## # A tibble: 1,128 x 5
##   pevpr valid_yr_count    lon    lat month
##   <dbl>          <dbl>  <dbl>  <dbl> <dbl>
## 1  1.02        1.19e-39   9.38   88.5    12
## 2  4.28        1.19e-39   9.38   86.7    12
## 3  3.03        9.18e-40   9.38   84.8    12
## # ... with 1,125 more rows
```

> 📖 Instead of extracting data for one longitude across latitudes, extract data across longitudes for one latitude near the Equator.

8.12   Remotely located data

Many of the functions described above accept an URL address in place of a file name. Consequently files can be read remotely without having to first download and save a copy in the local file system. This can be useful, especially when file names are generated within a script. However, one should avoid, especially in the case of servers open to public access, repeatedly downloading the same file as this unnecessarily increases network traffic and workload on the remote server. Because of this, our first example reads a small file from my own web site. See section 8.6 on page 299 for details on the use of these and other functions for reading text files.

```r
logger.df <-
    read.csv2(file = "http://r4photobiology.info/learnr/logger_1.txt",
              header = FALSE,
              col.names = c("time", "temperature"))
sapply(logger.df, class)
##        time temperature
## "character"   "numeric"

sapply(logger.df, mode)
##        time temperature
## "character"   "numeric"


logger.tb <-
    read_csv2(file = "http://r4photobiology.info/learnr/logger_1.txt",
              col_names = c("time", "temperature"))

## Using ',' as decimal and '.' as grouping mark. Use read_delim() for more control.
## Parsed with column specification:
## cols(
##   time = col_character(),
##   temperature = col_double()
## )

sapply(logger.tb, class)
##        time temperature
## "character"   "numeric"

sapply(logger.tb, mode)
##        time temperature
## "character"   "numeric"
```

While functions in package 'readr' support the use of URLs, those in packages 'readxl' and 'xlsx' do not. Consequently, we need to first download the file and save

a copy locally, that we can read as described in section 8.9.2 on page 312. Function
`download.file()` in the R 'utils' package can be used to download files using URLs. It
supports different modes such as binary or text, and write or append, and different
methods such as `"internal"`, `"wget"` and `"libcurl"` .

> ⚠ For portability, MS-Excel files should be downloaded in binary mode, set-
> ting `mode = "wb"`, which is required under MS-Windows.

```
download.file("http://r4photobiology.info/learnr/my-data.xlsx",
              "data/my-data-dwn.xlsx",
              mode = "wb")
```

Functions in package 'foreign', as well as those in package 'haven', support
URLs. See section 8.10 on page 316 for more information about importing this
kind of data into R.

```
remote_thiamin.df <-
  read.spss(file = "http://r4photobiology.info/learnr/thiamin.sav",
            to.data.frame = TRUE)
head(remote_thiamin.df)
##   THIAMIN CEREAL
## 1     5.2  wheat
## 2     4.5  wheat
## 3     6.0  wheat
## 4     6.1  wheat
## 5     6.7  wheat
## 6     5.8  wheat
```

```
remote_my_spss.tb <-
  read_sav(file = "http://r4photobiology.info/learnr/thiamin.sav")
remote_my_spss.tb
## # A tibble: 24 x 2
##   THIAMIN    CEREAL
##     <dbl> <dbl+lbl>
## 1     5.2 1 [wheat]
## 2     4.5 1 [wheat]
## 3     6   1 [wheat]
## # ... with 21 more rows
```

In this example we use a downloaded NetCDF file of long-term means for potential evapotranspiration from NOOA, the same used above in the 'ncdf4' example.
This is a moderately large file at 444 KB. In this case, we cannot directly open the
connection to the NetCDF file, and we first download it (commented out code, as
we have a local copy), and then we open the local file.

```
my.url <- paste("ftp://ftp.cdc.noaa.gov/Datasets/ncep.reanalysis.derived/",
                "surface_gauss/pevpr.sfc.mon.ltm.nc",
                sep = "")
#download.file(my.url,
#              mode = "wb",
#              destfile = "extdata/pevpr.sfc.mon.ltm.nc")
pet_ltm.nc <- nc_open("extdata/pevpr.sfc.mon.ltm.nc")
```

> ⚠ For portability, NetCDF files should be downloaded in binary mode, setting `mode = "wb"`, which is required under MS-Windows. .

8.13 Data acquisition from physical devices

Numerous modern data acquisition devices based on microcontrollers, including internet-of-things (IoT) devices, have servers (or daemons) that can be queried over a network connection to retrieve either real-time or logged data. Formats based on XML schemas or in JSON format are commonly used.

8.13.1 'jsonlite'

We give here a simple example using a module from the *YoctoPuce* (`http://www.yoctopuce.com/`) family using a software hub running locally. We retrieve logged data from a YoctoMeteo module.

> ⓘ This example needs setting the configuration of the YoctoPuce module beforehand. Fully reproducible examples, including configuration instructions, will be provided online.

Here we use function `fromJSON()` from package 'jsonlite' to retrieve logged data from one sensor.

```
hub.url <- "http://localhost:4444/"
Meteo01.df <-
    fromJSON(paste(hub.url, "byName/METEO01/dataLogger.json",
                   sep = ""), flatten = TRUE)
str(Meteo01.df, max.level = 2)
```

The minimum, mean, and maximum values for each logging interval need to be split from a single vector. We do this by indexing with a logical vector (recycled). The data returned is in long form, with quantity names and units also returned by the module, as well as the time.

```
Meteo01.df[["streams"]][[which(Meteo01.df$id == "temperature")]] %.>%
  as_tibble(x = .) %.>%
  dplyr::transmute(.data = .,
                   utc.time = as.POSIXct(utc, origin = "1970-01-01", tz = "UTC"),
                   t_min = unlist(val)[c(TRUE, FALSE, FALSE)],
                   t_mean = unlist(val)[c(FALSE, TRUE, FALSE)],
                   t_max = unlist(val)[c(FALSE, FALSE, TRUE)]) -> temperature.df

Meteo01.df[["streams"]][[which(Meteo01.df$id == "humidity")]] %.>%
  as_tibble(x = .) %.>%
```

```
dplyr::transmute(.data = .,
                 utc.time = as.POSIXct(utc, origin = "1970-01-01", tz = "UTC"),
                 hr_min = unlist(val)[c(TRUE, FALSE, FALSE)],
                 hr_mean = unlist(val)[c(FALSE, TRUE, FALSE)],
                 hr_max = unlist(val)[c(FALSE, FALSE, TRUE)]) -> humidity.df

full_join(temperature.df, humidity.df)
```

> ☕ Most YoctoPuce input modules have a built-in datalogger, and the stored data can also be downloaded as a csv file through a physical or virtual hub. As shown above, it is possible to control them through the HTML server in the physical or virtual hubs. Alternatively the R package 'reticulate' can be used to control YoctoPuce modules by means of the Python library giving access to their API.

8.14 Databases

One of the advantages of using databases is that subsets of cases and variables can be retrieved, even remotely, making it possible to work in R both locally and remotely with huge data sets. One should remember that R natively keeps whole objects in RAM, and consequently, available machine memory limits the size of data sets with which it is possible to work. Package 'dbplyr' provides the tools to work with data in databases using the same verbs as when using 'dplyr' with data stored in memory (RAM) (see chapter 6). This is an important subject, but extensive enough to be outside the scope of this book. We provide a few simple examples to show the very basics but interested readers should consult *R for Data Science* (Wickham and Grolemund 2017).

The additional steps compared to using 'dplyr' start with the need to establish a connection to a local or remote database. We will use R package 'RSQLite' to create a local temporary SQLite database. 'dbplyr' backends supporting other database systems are also available. We will use meteorological data from 'learnrbook' for this example.

```
library(dplyr)
con <- DBI::dbConnect(RSQLite::SQLite(), dbname = ":memory:")
copy_to(con, weather_wk_25_2019.tb, "weather",
        temporary = FALSE,
        indexes = list(
          c("month_name", "calendar_year", "solar_time"),
          "time",
          "sun_elevation",
          "was_sunny",
          "day_of_year",
          "month_of_year"
        )
```

```
)
weather.db <- tbl(con, "weather")
colnames(weather.db)
##  [1] "time"           "PAR_umol"        "PAR_diff_fr"   "global_watt"
##  [5] "day_of_year"    "month_of_year"   "month_name"    "calendar_year"
##  [9] "solar_time"     "sun_elevation"   "sun_azimuth"   "was_sunny"
## [13] "wind_speed"     "wind_direction"  "air_temp_C"    "air_RH"
## [17] "air_DP"         "air_pressure"    "red_umol"      "far_red_umol"
## [21] "red_far_red"

weather.db %.>%
  filter(., sun_elevation > 5) %.>%
  group_by(., day_of_year) %.>%
  summarise(., energy_wh = sum(global_watt, na.rm = TRUE) * 60 / 3600)
## # Source:    lazy query [?? x 2]
## # Database: sqlite 3.30.1 [:memory:]
##    day_of_year energy_wh
##          <dbl>     <dbl>
## 1          162     7500.
## 2          163     6660.
## 3          164     3958.
## # ... with more rows
```

> ☕ Package 'dbplyr' translates data pipes that use 'dplyr' syntax into SQL queries to databases, either local or remote. As long as there are no problems with the backend, the use of a database is almost transparent to the R user.

> ⓘ It is always good to clean up, and in the case of the book, the best way to test that the examples can be run in a "clean" system.
>
> ```
> unlink("./data", recursive = TRUE)
> unlink("./extdata", recursive = TRUE)
> ```

8.15 Further reading

Since this is the end of the book, I recommend as further reading the writings of Burns as they are full of insight. Having arrived at the end of *Learn R: As a Language* you should read *S Poetry* (Burns 1998) and *Tao Te Programming* (Burns 2012). If you want to never get caught unaware by R's idiosyncrasies, read also *The R Inferno* (Burns 2011).

Bibliography

Aho, A. V. and J. D. Ullman (1992). *Foundations of computer science*. Computer Science Press. ISBN: 0716782332.

Aiken, H., A. G. Oettinger, and T. C. Bartee (Aug. 1964). "Proposed automatic calculating machine". In: *IEEE Spectrum* 1.8, pp. 62–69. DOI: 10.1109/mspec.1964.6500770.

Becker, R. A. and J. M. Chambers (1984). *S: An Interactive Environment for Data Analysis and Graphics*. Chapman and Hall/CRC. ISBN: 0-534-03313-X (cit. on p. 3).

Becker, R. A., J. M. Chambers, and A. R. Wilks (1988). *The New S Language: A Programming Environment for Data Analysis and Graphics*. Chapman & Hall. ISBN: 0-534-09192-X (cit. on pp. 2, 3).

Boas, R. P. (1981). "Can we make mathematics intelligible?" In: *The American Mathematical Monthly* 88.10, pp. 727–731.

Burns, P. (1998). *S Poetry* (cit. on pp. 173, 326).

— (2011). *The R Inferno*. URL: http://www.burns-stat.com/pages/Tutor/R_inferno.pdf (visited on 07/27/2017) (cit. on p. 326).

— (2012). *Tao Te Programming*. Lulu. ISBN: 9781291130454 (cit. on p. 326).

Chambers, J. M. (2016). *Extending R*. The R Series. Chapman and Hall/CRC. ISBN: 1498775713 (cit. on pp. 3, 177).

Chang, W. (2018). *R Graphics Cookbook*. 2nd ed. O'Reilly UK Ltd. ISBN: 1491978600 (cit. on pp. 278, 291).

Cleveland, W. S. (1985). *The Elements of Graphing Data*. Wadsworth, Inc. ISBN: 978-0534037291 (cit. on p. 204).

Crawley, M. J. (2012). *The R Book*. Wiley, p. 1076. ISBN: 0470973927 (cit. on p. 161).

Dalgaard, P. (2008). *Introductory Statistics with R*. Springer, p. 380. ISBN: 0387790543 (cit. on p. 161).

Diez, D., M. Cetinkaya-Rundel, and C. D. Barr (2019). *OpenIntro Statistics*. 4th ed. 422 pp. URL: https://www.openintro.org/stat/os4.php (visited on 10/10/2019) (cit. on p. 161).

Eddelbuettel, D. (2013). *Seamless R and C++ Integration with Rcpp*. Springer, p. 248. ISBN: 1461468671 (cit. on p. 164).

Everitt, B. and T. Hothorn (2011). *An Introduction to Applied Multivariate Analysis with R*. Springer, p. 288. ISBN: 1441996494 (cit. on p. 161).

Everitt, B. S. and T. Hothorn (2009). *A Handbook of Statistical Analyses Using R*. 2nd ed. Chapman & Hall, p. 376. ISBN: 1420079336 (cit. on p. 161).

Faraway, J. J. (2004). *Linear Models with R*. Boca Raton, FL: Chapman & Hall/CRC, p. 240 (cit. on p. 161).

Faraway, J. J. (2006). *Extending the linear model with R: generalized linear, mixed effects and nonparametric regression models*. Chapman & Hall/CRC Taylor & Francis Group, p. 345. ISBN: 158488424X (cit. on p. 161).

Gandrud, C. (2015). *Reproducible Research with R and R Studio*. 2nd ed. Chapman & Hall/CRC The R Series). Chapman and Hall/CRC. 323 pp. ISBN: 1498715370 (cit. on pp. 9, 10).

Hamming, R. W. (Mar. 1, 1987). *Numerical Methods for Scientists and Engineers*. Dover Publications Inc. 752 pp. ISBN: 0486652416.

Hillebrand, J. and M. H. Nierhoff (2015). *Mastering RStudio: Develop, Communicate, and Collaborate with R*. Packt Publishing. 348 pp. ISBN: 9781783982554 (cit. on p. 8).

Holmes, S. and W. Huber (Mar. 1, 2019). *Modern Statistics for Modern Biology*. Cambridge University Press. 382 pp. ISBN: 1108705294 (cit. on p. 161).

Hughes, T. P. (2004). *American Genesis*. The University of Chicago Press. 530 pp. ISBN: 0226359271 (cit. on p. 92).

Johnson, K. A. and R. S. Goody (2011). "The Original Michaelis Constant: Translation of the 1913 Michaelis–Menten Paper". In: *Biochemistry* 50, pp. 8264–8269. DOI: 10.1021/bi201284u (cit. on p. 141).

Kernigham, B. W. and P. J. Plauger (1981). *Software Tools in Pascal*. Reading, Massachusetts: Addison-Wesley Publishing Company. 366 pp. (cit. on pp. 180, 187).

Kernighan, B. W. and R. Pike (1999). *The Practice of Programming*. Addison Wesley. 288 pp. ISBN: 020161586X (cit. on p. 15).

Knuth, D. E. (1984). "Literate programming". In: *The Computer Journal* 27.2, pp. 97–111 (cit. on pp. 9, 91).

Lamport, L. (1994). *LaTeX: a document preparation system*. English. 2nd ed. Reading: Addison-Wesley, p. 272. ISBN: 0-201-52983-1 (cit. on p. 91).

Leisch, F. (2002). "Dynamic generation of statistical reports using literate data analysis". In: *Proceedings in Computational Statistics*. Compstat 2002. Ed. by W. Härdle and B. Rönz. Heidelberg, Germany: Physika Verlag, pp. 575–580. ISBN: 3-7908-1517-9 (cit. on p. 9).

Lemon, J. (2020). *Kickstarting R*. URL: https://cran.r-project.org/doc/contrib/Lemon-kickstart/kr_intro.html (visited on 02/07/2020).

Loo, M. P. van der and E. de Jonge (2012). *Learning RStudio for R Statistical Computing*. 1st ed. Birmingham: Packt Publishing, p. 126. ISBN: 9781782160601 (cit. on p. 8).

Matloff, N. (2011). *The Art of R Programming: A Tour of Statistical Software Design*. No Starch Press, p. 400. ISBN: 1593273843 (cit. on pp. 86, 117, 179).

Murrell, P. (2011). *R Graphics*. 2nd ed. CRC Press, p. 546. ISBN: 1439831769 (cit. on p. 203).

— (2019). *R Graphics*. 3rd ed. Portland: CRC Press/Taylor & Francis. 423 pp. ISBN: 1498789056 (cit. on pp. 83, 291).

Newham, C. and B. Rosenblatt (June 1, 2005). *Learning the bash Shell*. O'Reilly UK Ltd. 352 pp. ISBN: 0596009658 (cit. on p. 15).

Peng, R. D. (2016). *R Programming for Data Science*. Leanpub. 182 pp. URL: https://leanpub.com/rprogramming (visited on 07/31/2019) (cit. on pp. 86, 181).

Pinheiro, J. C. and D. M. Bates (2000). *Mixed-Effects Models in S and S-Plus*. New York: Springer (cit. on pp. 161, 164).

Sarkar, D. (2008). *Lattice: Multivariate Data Visualization with R*. 1st ed. Springer, p. 268. ISBN: 0387759689 (cit. on pp. 83, 164, 203).

Smith, H. F. (1957). "Interpretation of adjusted treatment means and regressions in analysis of covariance". In: *Biometrics* 13, pp. 281–308 (cit. on p. 138).

Tufte, E. R. (1983). *The Visual Display of Quantitative Information*. Cheshire, CT: Graphics Press. 197 pp. ISBN: 0-9613921-0-X (cit. on p. 241).

Venables, W. N. and B. D. Ripley (2002). *Modern Applied Statistics with S*. 4th. New York: Springer. ISBN: 0-387-95457-0 (cit. on p. 161).

Wickham, H. (2015). *R Packages*. O'Reilly Media. ISBN: 9781491910542 (cit. on pp. 164, 177).

Wickham, H. (2019). *Advanced R*. 2nd ed. Taylor & Francis Inc. 588 pp. ISBN: 0815384572 (cit. on pp. 117, 177).

Wickham, H. and G. Grolemund (2017). *R for Data Science*. O'Reilly UK Ltd. ISBN: 1491910399 (cit. on pp. 181, 201, 325).

Wickham, H. and C. Sievert (2016). *ggplot2: Elegant Graphics for Data Analysis*. 2nd ed. Springer. XVI + 260. ISBN: 978-3-319-24277-4 (cit. on pp. 164, 203, 278, 291).

Wood, S. N. (2017). *Generalized Additive Models*. Taylor & Francis Inc. 476 pp. ISBN: 1498728332 (cit. on p. 161).

Xie, Y. (2013). *Dynamic Documents with R and knitr*. The R Series. Chapman and Hall/CRC, p. 216. ISBN: 1482203537 (cit. on pp. 9, 10, 91).

— (2016). *bookdown: Authoring Books and Technical Documents with R Markdown*. Chapman and Hall/CRC. ISBN: 9781138700109 (cit. on p. 91).

Xie, Y., J. J. Allaire, and G. Grolemund (2018). *R Markdown*. Chapman and Hall/CRC. 304 pp. ISBN: 1138359335 (cit. on p. 91).

Zachry, M. and C. Thralls (Oct. 2004). "An Interview with Edward R. Tufte". In: *Technical Communication Quarterly* 13.4, pp. 447–462. DOI: 10.1207/s15427625tcq1304_5.

Zuur, A. F., E. N. Ieno, and E. Meesters (2009). *A Beginner's Guide to R*. 1st ed. Springer, p. 236. ISBN: 0387938362 (cit. on p. 161).

General index

Index of R names by category

Alphabetic index of R names

||, 29

*, 18, 35
+, 18, 28, 281
-, 18, 28
->, 20
-Inf, 25, 34
.Machine$double.eps, 34
.Machine$double.max, 34
.Machine$double.min, 34
.Machine$double.neg.eps, 34
.Machine$integer.max, 35
/, 18, 281
:, 23
<, 31
<-, 19, 20, 48, 73
<<-, 167
<=, 31
=, 20
==, 31, 197
>, 31
>=, 31
[,], 197
[ ], 64, 73
[[ ]], 64
[[]], 62, 67, 70, 71
[], 70, 76
$, 68, 70, 71
%*%, 56
%.>%, 189, 194
%/%, 27
%<>%, 189
%>%, 188, 189
%T>%, 189
%%, 27
%in%, 37, 39
&, 29
&&, 29
^, 35
|, 29

abs(), 28, 36
aes(), 215, 231
aggregate(), 195
AIC(), 131
all(), 30
annotate(), 235, 270, 272
annotation_custom(), 236, 270
anova(), 116, 131, 132, 135, 137, 138
anti_join(), 200
any(), 30
aov(), 135, 153
append(), 22, 63
apply(), 107, 108, 111, 112
arrange(), 194
array, 51
array(), 54
as.character(), 42, 58
as.data.frame(), 185
as.formula(), 148
as.integer(), 43
as.logical(), 42
as.matrix(), 51
as.numeric(), 42, 43, 58
as.ts(), 151
as.vector(), 55
as_tibble(), 183
assign(), 113, 114, 167
attach(), 72-75
attr(), 77
attr()<-, 77
attributes(), 77, 127, 318
austres, 151

basename(), 296
BIC(), 131
biplot(), 156

T - #0057 - 171024 - C364 - 254/178/17 - PB - 9780367182533 - Gloss Lamination